Frederik Wegelin

Magnetization Dynamics in Spin Valves

Frederik Wegelin

Magnetization Dynamics in Spin Valves

A study using Time Resolved Photo-Emission Electron Microscopy in combination with X-ray Circular Dichroism (TR-PEEM-XMCD)

Südwestdeutscher Verlag für Hochschulschriften

Impressum / Imprint
Bibliografische Information der Deutschen Nationalbibliothek: Die Deutsche Nationalbibliothek verzeichnet diese Publikation in der Deutschen Nationalbibliografie; detaillierte bibliografische Daten sind im Internet über http://dnb.d-nb.de abrufbar.
Alle in diesem Buch genannten Marken und Produktnamen unterliegen warenzeichen-, marken- oder patentrechtlichem Schutz bzw. sind Warenzeichen oder eingetragene Warenzeichen der jeweiligen Inhaber. Die Wiedergabe von Marken, Produktnamen, Gebrauchsnamen, Handelsnamen, Warenbezeichnungen u.s.w. in diesem Werk berechtigt auch ohne besondere Kennzeichnung nicht zu der Annahme, dass solche Namen im Sinne der Warenzeichen- und Markenschutzgesetzgebung als frei zu betrachten wären und daher von jedermann benutzt werden dürften.

Bibliographic information published by the Deutsche Nationalbibliothek: The Deutsche Nationalbibliothek lists this publication in the Deutsche Nationalbibliografie; detailed bibliographic data are available in the Internet at http://dnb.d-nb.de.
Any brand names and product names mentioned in this book are subject to trademark, brand or patent protection and are trademarks or registered trademarks of their respective holders. The use of brand names, product names, common names, trade names, product descriptions etc. even without a particular marking in this work is in no way to be construed to mean that such names may be regarded as unrestricted in respect of trademark and brand protection legislation and could thus be used by anyone.

Verlag / Publisher:
Südwestdeutscher Verlag für Hochschulschriften
ist ein Imprint der / is a trademark of
OmniScriptum GmbH & Co. KG
Heinrich-Böcking-Str. 6-8, 66121 Saarbrücken, Deutschland / Germany
Email: info@svh-verlag.de

Herstellung: siehe letzte Seite /
Printed at: see last page
ISBN: 978-3-8381-1573-3

Zugl. / Approved by: Mainz, Johannes-Gutenberg Universität, Diss., 2010

Copyright © 2010 OmniScriptum GmbH & Co. KG
Alle Rechte vorbehalten. / All rights reserved. Saarbrücken 2010

Betreuer. Prof. Dr. G Schönhense

2. Berichterstatter: Prof. Dr. T. Doll

Tag der mündlichen Prüfung: 4.2.2010

Y EL MUNDO SE PARO PARA UN MOMENTO…

…Y NACIÓ UNA ESTRELLA

POR JULIAN-FREDERIK

Kurzfassung

Schlüsseltechnologien wie magnetoresistive Sensoren oder das MRAM (Magnetic Random Access Memory) erfordern reproduzierbare magnetische Schaltvorgänge zwischen remanenten Zuständen. In modernen magnetischen Speichern erreicht die Geschwindigkeit solcher Prozesse die Zeitskala der gyromagnetischen Präzession. Die Landau-Lifschitz-Glibert-Gleichung (LLG) beschreibt magnetisch angeregte Zustände in Form von Eigenmoden und Dämpfung in lateral begrenzten dünnen Schichten.

Forschungsobjekte dieser Arbeit sind u.a. hochentwickelte anti-parallel gepinnte synthetische Spinventile, wie sie als GMR-Leseelemente (Giant MagnetoResistive) in heutigen Magnetspeicherplatten zum Einsatz kommen. Darin ist eine ferromagnetische Schicht hoher Koerzitivität mittels Austauschkopplung an einen Antiferromagneten gekoppelt. Eine zweite, durch einen nichtmagnetischen Spacer getrennte ferromagnetische Schicht, richtet sich bei entsprechender Spacerdicke antiparallel zu dieser aus. Eine dritte, wiederum antiparallel zur zweiten ausgerichtete NiFe-Schicht niedriger Koerzitivität (der Freelayer), wirkt als sensierende Schicht, in dem sich, abhängig vom äußeren Magnetfeld, der Widerstand im Schichtstapel ändert. In dieser Arbeit wird mittels elementspezifischer, zeitaufgelöster Photoemissions-Elektronenmikroskopie (TR-PEEM), die Magnetisierung des Freelayers isoliert vom Rest des Schichtstapels abgebildet. Genutzt wird dazu der Röntgenzirkulardichroismus (XMCD).

Die untersuchten Spinventile (typisch $\Delta R/R$ = 15%) und wurden mittels Dünnfilmtechnik auf den pulsführenden Leiter eines koplanaren Wellenleiters aufgebracht. Dabei wurden Geometrie (Rechtecke, Ellipsen, Ringe), Dimension (im Bereich weniger Mikrometer) und Orientierung zum Feldpuls variiert. Um magnetische Schaltvorgänge im Gigahertz-Regime überhaupt untersuchen zu können, mussten zum einen ultra-kurze Röntgenpulse am Synchrotron BESSY II erzeugt werden (low α-mode) und zum anderen die Wellenleitertechnik zur hochfrequenten elektromagnetischen Anregung (FWHM typisch 100 ps) optimiert werden. Orts- u. Zeitauflösung konnten auf d = 100 nm bzw. Δt = 15 ps reduziert werden.

Es wird gezeigt, dass die Magnetisierungsdynamik des Freelayers in einem GMR-Spinventil nicht der erwarteten kohärenten Rotation entspricht. Vielmehr ist die dynamische Antwort des Systems auf den Feldpuls eine Überlagerung aus einer homogenen kritisch gedämpften Präzession und Spinwellenmoden höhere Ordnung. Für eine quadratische Struktur wird eine Spinwellenmode beobachtet, deren Periode 600 ps (1.7 GHz) beträgt. Der Dämpfungskoeffizient erweist sich augenscheinlich zunächst als unabhängig von der Form des Spinventils, wodurch das Modell homogener Rotation und Dämpfung gestützt würde. Erst die Differenzbildung der magnetischen Rotation zwischen Mitte und Randbereich der Struktur macht die Spinwellenmode höhere Ordnung sichtbar, deren Beitrag zur Energiedissipation zu einer erhöhten effektiven Dämpfung (α = 0.01) führt. Dämpfung und magnetische Schaltvorgänge in Spinventilen hängen somit von der Geometrie ab. Mikromagnetische Simulationen reproduzieren die beobachteten Spinwellenmoden.

Neben dem Kurzzeitverhalten der Magnetisierung von Spinventilen wurden einlagige Permalloy-Schichten mit Dicken zwischen 3 und 40 nm untersucht. Die Phasengeschwindigkeit einer Spinwelle konnte in einer 3 nm dicken Ellipse mit 8.100 m/s ermittelt werden. Für eine rechteckige Struktur mit Landau-Domänen beträgt die gemessene Geschwindigkeit der Verschiebung einer 90°-Néel-Wand durch den Feldpuls, 15.000 m/s.

Abstract

Key technology applications like magnetoresistive sensors or the Magnetic Random Access Memory (MRAM) require reproducible magnetic switching mechanisms. i.e. predefined remanent states. At the same time advanced magnetic recording schemes push the magnetic switching time into the gyromagnetic regime. According to the Landau-Lifschitz-Gilbert formalism, relevant questions herein are associated with magnetic excitations (eigenmodes) and damping processes in confined magnetic thin film structures.

Objects of study in this thesis are antiparallel pinned synthetic spin valves as they are extensively used as read heads in today's magnetic storage devices. In such devices a ferromagnetic layer of high coercivity is stabilized via an exchange bias field by an antiferromagnet. A second hard magnetic layer, separated by a non-magnetic spacer of defined thickness, aligns antiparallel to the first. The orientation of the magnetization vector in the third ferromagnetic NiFe layer of low coercivity - the freelayer - is then sensed by the Giant MagnetoResistance (GMR) effect. This thesis reports results of element specific Time Resolved Photo-Emission Electron Microscopy (TR-PEEM) to image the magnetization dynamics of the free layer alone via X-ray Circular Dichroism (XMCD) at the Ni-L_3 X-ray absorption edge.

The ferromagnetic systems, i.e. micron-sized spin valve stacks of typically $\Delta R/R = 15\%$ and Permalloy single layers, were deposited onto the pulse leading centre stripe of coplanar wave guides, built in thin film wafer technology. The ferromagnetic platelets have been applied with varying geometry (rectangles, ellipses and squares), lateral dimension (in the range of several micrometers) and orientation to the magnetic field pulse to study the magnetization behaviour in dependence of these magnitudes. The observation of magnetic switching processes in the gigahertz range became only possible due to the joined effort of producing ultra-short X-ray pulses at the synchrotron source BESSY II (operated in the so-called low-α mode) and optimizing the wave guide design of the samples for high frequency electromagnetic excitation (FWHM typically several 100 ps). Space and time resolution of the experiment could be reduced to d = 100 nm and Δt = 15 ps, respectively.

In conclusion, it could be shown that the magnetization dynamics of the free layer of a synthetic GMR spin valve stack deviates significantly from a simple phase coherent rotation. In fact, the dynamic response of the free layer is a superposition of an averaged critically damped precessional motion and localized higher order spin wave modes. In a square platelet a standing spin wave with a period of 600 ps (1.7 GHz) was observed. At a first glance, the damping coefficient was found to be independent of the shape of the spin-valve element, thus favouring the model of homogeneous rotation and damping. Only by building the difference in the magnetic rotation between the central region and the outer rim of the platelet, the spin wave becomes visible. As they provide an additional efficient channel for energy dissipation, spin waves contribute to a higher effective damping coefficient (α = 0.01). Damping and magnetic switching behaviour in spin valves thus depend on the geometry of the element. Micromagnetic simulations reproduce the observed higher-order spin wave mode.

Besides the short-run behaviour of the magnetization of spin valves Permalloy single layers with thicknesses ranging from 3 to 40 nm have been studied. The phase velocity of a spin wave in a 3 nm thick ellipse could be determined to 8.100 m/s. In a rectangular structure exhibiting a Landau-Lifschitz like domain pattern, the speed of the field pulse induced displacement of a 90°-Néel wall has been determined to 15.000 m/s.

Content

CHAPTER 1 INTRODUCTION — 9

CHAPTER 2 THEORY OF FAST REMAGNETIZATION PROCESSES — 14
2.1. Formation of Magnetic Domains and Domain Walls — 15
2.2. The Landau-Lifschitz-Gilbert Formalism — 18
2.3. Magnetization Dynamics in extended Structures and coupled Layers — 22

CHAPTER 3 EXPERIMENTAL — 25
3.1. Stroboscopic Imaging using the Time Structure of Synchrotron Radiation — 26
3.2. Time-resolved XMCD-PEEM — 29

CHAPTER 4 PREPARATION OF MICRO STRIPLINE SAMPLES — 32
4.1. Preparation of Micro Strip Line samples with Spin-Valves — 35
    Design Optimization — 35
    Device Fabrication — 38
4.2. Preparation of Single Crystalline Co Platelets — 48
4.3. Ultra-thin Permalloy Rings and Ellipses — 55
4.4. Thicker Permalloy Platelets — 57

CHAPTER 5 RESULTS — 58
5.1. Magnetization Dynamics in Microscopic Spin Valves — 59
5.2. Epitaxial Co Platelets on a Single-Crystalline Mo Strip Line — 75
5.3. Propagating Magnetic Eigenmode in Ultra-Thin Py Ellipse — 78
5.4. Magnetization Dynamics of Ultra-Thin Py Rings — 81
5.5. Dynamic Response of 90° Néel Domain Walls — 84

CHAPTER 6 CONCLUSIONS AND OUTLOOK — 92

CHAPTER 7 ATTACHMENTS — 96
7.1. List of Publications — 97
7.2. Abbreviations — 99
7.3. References — 100

Chapter 1

Introduction

Key technology applications like the magnetic random access memory (MRAM) or magnetoresistive sensors require reproducible switching mechanisms. i.e. predefined remanent states. The switching process itself has to be reproducible. Thus, understanding the fundamental physics of the magnetic behaviour of two-dimensional micron sized ferromagnetic elements is inevitable to control the switching process. Furthermore, advanced magnetic recording schemes and spintronics push the magnetic switching time into the gyromagnetic regime. Ultrafast magnetization excitations in soft magnetic microstructures thus recently attracted particular attention[1,2,3,4,5,6]. New switching concepts involving the spin transfer torque[7,8] also rely on gyromagnetic processes. For microscopic elements with small magnetic anisotropy and a well-defined shape, the high-frequency behaviour is governed by confined spin wave eigenmodes[1,2,9].

Relevant questions are associated with magnetic excitations (eigenmodes) and damping processes in confined magnetic structures[10]. One the one hand, there is some work in the frequency domain, i.e. high-frequency spectroscopy (ferromagnetic resonance) in nanoscale dots[11] and references therein. On the other hand, work in the time domain is sparse because it requires an extremely high time resolution to observe processes in the gigahertz range directly.

Magnetic modes have been studied recently by Park et al. using time-resolved Kerr-microscopy[5,12], by XMCD–PEEM[1,3,13] and by transmission X-ray microscopy[14]. In addition to spinwave modes, transient spatio-temporal features like the formation of transient domain walls and transient vortices as well as the emergence of stripe-like domain patterns (blocking patterns) have been observed by Kerr microscopy[15] and PEEM[16].

Magnetic spin valves represent very important functional structures in modern magnetism. Today, spin valves are extensively used as read heads in magnetic storage devices. Their functionality depends crucially on the interplay of magnetic coupling phenomena. In its simplest version, a spin valve is composed of two ferromagnetic (FM) layers separated by a nonmagnetic (NM) spacer layer mediating a usually antiferromagnetic indirect exchange coupling[17,18] which determines the magnetic configuration of the layer stack. In a more refined approach, the magnetization in one of the FM layers (hard layer) is additionally stabilized by a strong coupling (exchange biasing) to an antiferromagnet. The orientation of the magnetization vector in the other FM layer — the free layer — is then sensed by the giant magneto resistance (GMR) effect[18,19,20]. In more complex systems, further coupling mechanisms such as orange peel or edge coupling may take place[21].

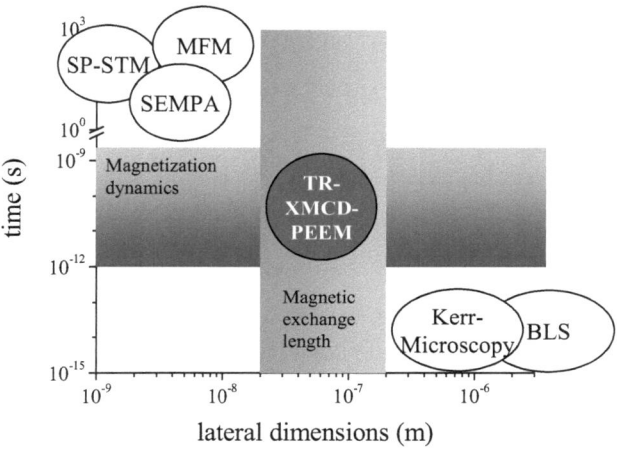

Fig. 1: Comparison of different magnetic imaging techniques in the landscape of time scale and lateral dimension. A time resolution of 15 ps, a lateral resolution of approx. 100 nm and several nanometers of depth sensitivity make Time Resolved Photo Emission Electron Microscopy (TR-PEEM) unique and the preferred investigation method for imaging spin dynamics (figure courtesy C.M. Schneider[48]).

Micron-sized spin valves are extremely interesting structures from a fundamental point of view, as they provide a unique access to the interplay between different types of magnetic coupling in both static and dynamic experiments[22]. Owing to its unique time structure, circularly polarized synchrotron radiation has proven to be an excellent tool for high-speed imaging of such ultrafast magnetization processes. In full-field imaging photoemission electron microscopy (PEEM) combined with XMCD (X-ray magnetic circular dichroism) the contrast is very high and provides element selectivity[23]. Time resolution can be as good as 15 ps in special bunch-compression modes of the storage ring[24] and the depth sensitivity of several nanometres can give access to buried layers in ultrathin-film systems[25].

These characteristics make XMCD–PEEM a powerful method in an interesting region of the landscape time scale versus lateral dimension (see Fig. 1).

Other imaging techniques like magnetic force microscopy (MFM), secondary electron microscopy with polarization analysis (SEMPA) and spin-polarized scanning tunnelling microscopy (SP-STM) provide higher lateral resolution but insufficient temporal resolution. Optical Kerr microscopy and Brillouin light scattering (BLS), on contrary, yield very high time resolution but insufficient spatial resolution.

In the quasi-static regime, the magnetization pattern almost instantaneously follows the slowly varying applied field via nucleation and motion of domain walls and vortices, magnetization rotation, and Barkhausen jumps. Such processes have been studied in detail using quasi-static Kerr-microscopy[26]. Soft magnetic elements minimise their stray field energy by forming inhomogeneous magnetization structures (e.g. Landau flux-closure patterns). Magnetic eigenmodes, i.e. the so-called wall modes, vortex modes and normal or centre modes are not excited in this regime. For fast field pulses the frequency spectrum of the magnetic pulse contains significant Fourier components in the range of the eigenmode frequencies (typically 0.5 GHz to several GHz). Then these modes are excited and their characteristics will govern the dynamic magnetic response of the system.

In the present work, stroboscopic XMCD–PEEM exploiting the time structure of synchrotron radiation is used for probing fast switching processes, transient magnetic states as well as standing and propagating spin waves in a variety of different structures. These comprise ring shaped and elliptical structures of ultra-thin Permalloy, samples consisting of a single crystalline Co layer with high uniaxial anisotropy and the free layer investigation of GMR spin valves shaped into rectangles, squares and ellipse. The spin valve samples have been fabricated by industry standards using the thin film deposition, photolithography and ion beam facilities of Sensitec GmbH, Mainz. The Py and Co samples were manufactured by the group of C.M. Schneider, FZ Jülich.

It will be shown that although the averaged magnetization vector reacts on external field pulses according to a single-spin model with critical damping, local spin wave modes are excited depending on the shape of the spin valve structure. As spin waves provide an additional channel for energy dissipation, they contribute to a higher effective damping.

At a first glance, the fact that the damping coefficient is apparently independent of the shape favours the model of a nonlocal magnetization damping. Taking spin wave modes in account, damping and magnetic switching behaviour in spin valves depends on the element's geometry.

Chapter 2

Theory of fast Remagnetization Processes

2.1. Formation of Magnetic Domains and Domain Walls

Softmagnetic two-dimensional particles exceeding the critical expansion of the magnetic exchange length form domain patterns. The domain configuration is stable at the energetic minimum. The total energy is the sum of energy terms: Zeemann energy, demagnetizing field, exchange energy and several anisotropy terms (like crystal anisotropy, shape anisotropy or magnetostriction-induced anisotropy). Demagnetizing energy bases on the classical long ranged dipole-dipole interaction, whereas the exchange energy stems from the short-range interaction of next neighbours. The magnetization within a domain is thus expected to be homogeneous and the extension of domain walls (Néel and Bloch walls) small against the size of the domains.

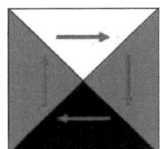
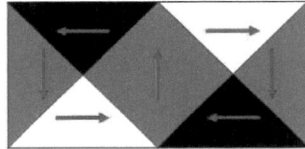

Fig. 2: Examples of Landau-Lifschitz patterns with vanishing stray field in micron sized two-dimensional ferromagnets as expected for a single layer of Permalloy of several nanometers thickness. For every domain the direction of the magnetic field vector is indicated by a red arrow.

Examples of Landau-Lifschitz patterns with vanishing stray field are shown in Fig. 2. In the junction of the four magnetization directions, the so called vortex, the magnetization points out of the plane. With decreasing particle size or increasing shape anisotropy the demagnetization field is reduced to a limit, at which the energy for domain wall generation cannot be afforded.

Due to the pole avoidance principle[10], which prevents the magnetization from rotating freely close to the platelet rim, the magnetization aligns along the easy axis and edge domains occur. For rectangular particles, these domains can align in two different ways, the so-called "S" state and the "C" state (see Fig. 3).

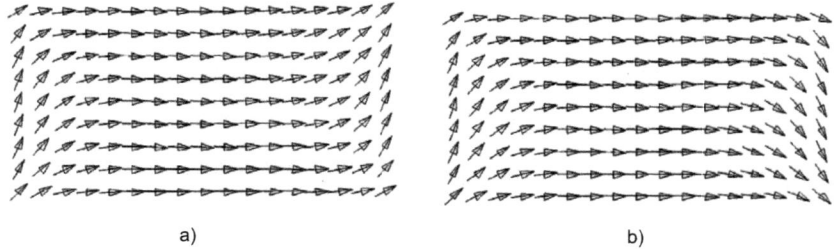

Fig. 3: Formation of edge domains in the "S" state (a) and the "C" state (b) as simulated by Zheng et al.[27]

This work will confirm that these residual edge domains do not participate in fast magnetization processes.

Ring shaped soft-magnetic thin film structures are promising scientific objects[28] for they have two well defined and reproducible magnetic states (see Fig. 4). In the vortex state the magnetic flux follows the ring structure forming a closed loop. The vortex state also represents the energetic minimum (for vanishing anisotropy) since no stray field energy is stored in the system (see Fig. 4a). The onion state on the other hand features two 180° head-to-head domains and a significant stray field (see Fig. 4b).

Whereas Py rings in onion state have been extensively studied statically, time resolved simulations have been presented by Kläui et al. from University of Konstanz[29] (see Fig. 4d). In this work ultra-thin rings of 3 nm thick permalloy in the onion state are investigated. Subject of investigation is the magnetic response of the system to a pulsed magnetic excitation.

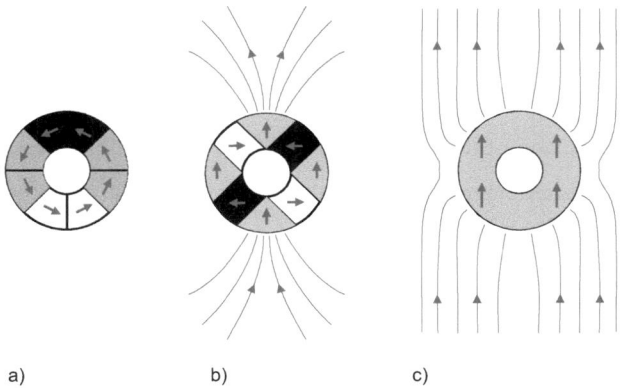

Fig. 4: Magnetic domain configuration of a Permalloy ring. a) The vortex state, manifested by a closed magnetic flux without stray field, b) the onion state with 180° head-to-head domains and c) ring magnetically saturated by an external field applied in vertical direction. For all samples, the red arrows indicate the direction of magnetization vector. The grey value schematically indicates the contrast in XMCD-PEEM for photon impact from left to right (see chapter 3, Fig. 13)

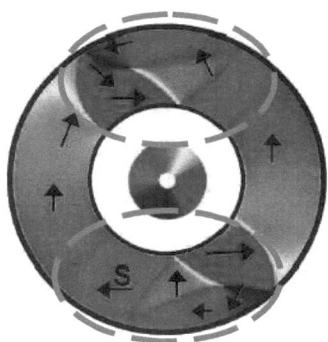

Fig. 5: Simulation of Py Ring in onion state with two 180° head-to-head domains marked by dashed lines[29].

In anti-parallel pinned synthetic spin valves, as investigated in this thesis, the magnetization within the free layer is expected to form a single-domain state. Domain wall and vortex formation is expected to be suppressed due to the high exchange coupling field biasing the softmagnetic free layer.

2.2. The Landau-Lifschitz-Gilbert Formalism

As the underlying formalism to magnetization dynamics is the Landau-Lifschitz-Gilbert equation, a brief review seems appropriate to understand the behaviour of the investigated elements. In a ferromagnetic solid with reduced dimension, a macro spin model can best be understood assuming that the temporal behaviour of all micro spins would be in coherence and picturing a submicron sized ferromagnetic particle with finite coercivity. If the particle size is small enough, the atomic magnetization vectors can be considered as quasi unidirectional.

The quantum mechanical origin of magnetic precession is the Schrödinger equation

$$i\hbar \frac{d}{dt}\langle \mathbf{S} \rangle_{(t)} = \langle \mathbf{S}, H_{(t)} \rangle \qquad (1)$$

With the Zeeman term of the Hamiltonian in vacuum

$$H = -\frac{g\mu_B}{\hbar} \mathbf{S}\mathbf{B}, \qquad \mathbf{B} = \mu_0 \mathbf{H}.$$

Here, μ_B is Bohr's magneton und g the gyromagnetic factor. For the temporal evolution of the mean value of the spin operator holds

$$\frac{d}{dt}\langle \mathbf{S} \rangle_{(t)} = \frac{g\mu_B}{\hbar} \left(\langle \mathbf{S} \rangle_{(t)} \times \mathbf{B}_{(t)} \right) \qquad (2).$$

Employing the relation between the magnetic dipole moment and the spin operator

$$\mathbf{M} = \frac{g\mu_B}{\hbar} \langle \mathbf{S} \rangle, \qquad (3)$$

the equation of motion (first order) for the magnetic dipole moment can be expressed as

$$\frac{d}{dt}\mathbf{M} = \frac{g\mu_B}{\hbar}\left[\mathbf{M}_{(t)} \times \mathbf{B}_{(t)}\right] \qquad (4)$$

The magnetization of the total dipole moment per unit volume will be

$$\mathbf{M} = \frac{\sum M(t)}{\text{unit.vol.}} \quad \text{and with} \quad \gamma_0 = \mu_0 \frac{g|\mu_B|}{\hbar}$$

follows

$$\frac{d}{dt}\mathbf{M}_{(t)} = -\gamma_0\left[\mathbf{M}_{(t)} \times \mathbf{H}_{(t)}\right]. \qquad (5)$$

Despite the factor γ_0 the equation of motion for a magnetic dipole moment is identical to it's classical equivalent, the relation between angular momentum and torque.

Damping of the precessional movement can be introduced into the equation of motion assuming the effective field

$$\mathbf{H}_{\textit{eff}} = \mathbf{H} - \alpha\frac{1}{\gamma_0 M_s}\frac{d\mathbf{M}}{dt},$$

where $H = H_{ext} + H_{aniso} + H_{demag}$. The demagnetization field H_{demag}, the sum of all anisotropy terms H_{aniso} and the external field H_{ext} constitute to H.

The Landau-Lifschitz-Gilbert equation[10, 30] (LLG) than concludes as

$$\frac{d}{dt}\mathbf{M}_{(t)} = -\gamma_0 [\mathbf{M}_{(t)} \times \mathbf{H}_{(t)}] + \frac{\alpha}{M_s}\left[\mathbf{M}_{(t)} \times \frac{d\mathbf{M}_{(t)}}{dt}\right]. \qquad (6)$$

The LLG bridges quantum and classical mechanics due to it's origin from Schrödinger's equation on one hand and it's classical appearance on the other. The observable $M_{(x,t)}$ is experimentally accessible in ferromagnetic layers with reduced dimension. At the same time the macro spin $M_{(x,t)}$ represents the collective phenomenon of quantum mechanical spin precession.

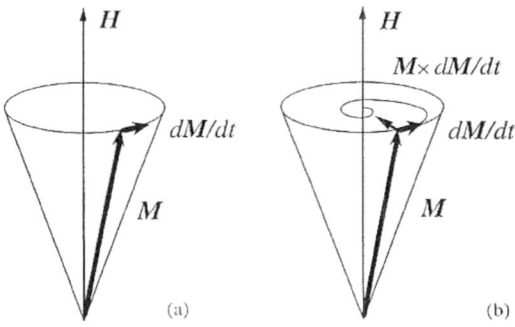

Fig. 6: The magnetization vector *M* performs a gyroscopic motion around the vector of the external magnetic field *H* without damping a). After introducing the Gilbert term *M* experiences a damping b)[10].

The model is equivalent to a macrospin reacting on an effective magnetic field as sketched in Fig. 6. The effective field H_{eff} contains all coupling contributions and exerts a torque on *M*, which initiates its precessional motion, if the Fourier spectrum of the exciting external field pulse comprises significant components of the precessional eigenfrequency of the system[9]. This also holds for a spin torque or a photon angular momentum transfer.

Due to the pulsed nature of these excitations, their frequency spectrum usually contains several eigenfrequencies of the system.

Weekly damped precession of spins induces the so-called ringing of magnetization, which can persist up to several nanoseconds. For most industrial applications this is an undesired effect and high damping coefficients are preferred.

2.3. Magnetization Dynamics in extended Structures and coupled Layers

Spin waves as subject of scientific research cover magnetic phenomena reaching from long time scale into the femtosecond time regime. Whereas lower frequencies of some kilohertz cover phenomena like domain wall motion, coherent rotational processes or phenomena of ferromagnetic resonance as depicted in Fig. 8 take place with some gigahertz[10]. Fig. 7 gives an overview to relevant magnetic phenomena associated to the order of magnitude in time scale. In ferromagnetic thin films, the dynamics of magnetic behaviour is limited by the time needed to equilibrate energetically the electron gas and the spin bath.

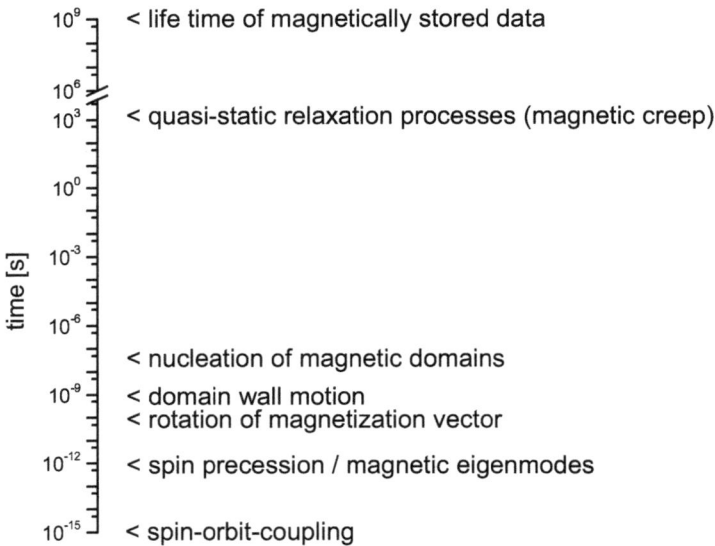

Fig. 7: Dynamic magnetic effects associated to coarse time scale. While magnetic flipping by rotation of the magnetization vector efforts picoseconds, spin precession takes place in femtoseconds. Quasi-static magnetic relaxation can last minutes or even several hours.

The intrinsic eigenspectrum of a magnetic system is represented by it's spin wave spectrum which is conventionally investigated with spectrometric methods like lateral and time resolved Brillouin scattering (BLS)[31] or ferromagnetic resonance spectroscopy[32]. The lateral resolution of BLS techniques for example is limited by the diameter of the probing laser and adds up to 30 to 50 µm[10]. GMR spin valves have also been subject to investigation with methods like FMR[33].

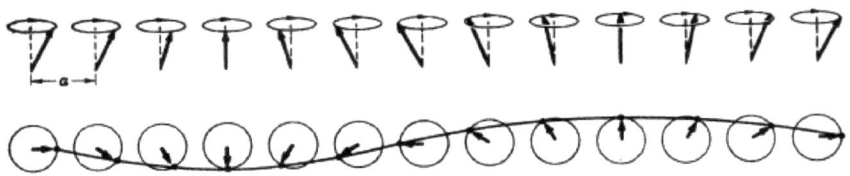

Fig. 8: Model of a standing transversal spin wave in ferromagnetic material. The magnetization vector rotates with respect to the lateral position. The boundary conditions cause the standing spin wave to appear as a multiple of the period[34].

Spin valve elements constitute of a coupled layer system (e.g. see inset in Fig. 9)[35]. The quasi-static behaviour is characterized by the giant magnetoresistive (GMR) effect. A typical magnetoresistance measurement for one the spin valves investigated in this thesis is shown in Fig. 9. Despite their lateral extension, the magnetization reversal in such spin-valve elements was successfully described by a coherent precessional path[36], assuming a global magnetization value and a high damping coefficient as depicted in Fig. 6. This finding is surprising, considering the fact that the dynamical behaviour of the local magnetization driven by the local effective field H_{eff} could influence the response of a spin-valve sensor and the damping coefficient. Besides the shape-induced demagnetization field, H_{eff} contains contributions from the correlated roughness at the FM/NM interfaces[37] and stray fields from inhomogeneously magnetized regions in one of the layers[38,39,40]. An inhomogeneous interlayer thickness may also cause a laterally varying indirect coupling field.

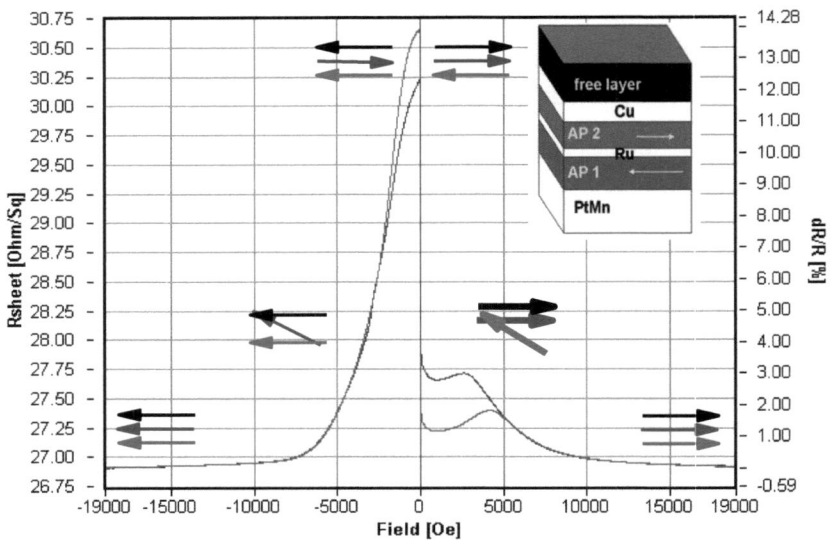

Fig. 9: Measured transfer curve of a typical GMR spin valve with $\Delta R/R \approx 15\%$. Inset: Schematic spin valve stack with Ruthenium spacer between the two hard magnetic ferromagnets and Cu spacer between hard magnetic layer and free layer (own measurement).

Finally, the dynamic magnetization reversal in the free layer may affect the magnetization configuration in the hard layer[41] (see Fig. 9). The free layer in spin valves was found to show an increased damping coefficient[39,42] which was attributed to a nonlocal spin pumping model[43,44]. A competing damping process, however, is given by the local excitation of additional short wavelength and high-frequency spin wave modes[45,46]. In general, a local variation of the precessional magnetization motion will lead to unwanted magnetically induced noise in the response of a spin valve[42] or any other fast-switching magnetic structure[47]. Therefore, the understanding of the local magnetization dynamics in complex layer stacks is extremely important, as the dynamics may be determined in a complicated way by the various magnetic coupling effects in a thin film structure[46]. The control of the damping coefficient α holds the key for optimized magnetic switching procedures[42].

Chapter 3

Experimental

3.1. Stroboscopic Imaging using the Time Structure of Synchrotron Radiation

The time-resolved experiments were performed with stroboscopic illumination of the sample with circular polarized (P_{circ}) x-ray pulses at the Ni L$_3$ absorption edge produced by electron bunches in the synchrotron ring (t_{FWHM} = 3 ps, low-α mode at BESSY II, Berlin) with a repetition rate of 500 MHz. The magnetic response of the element is tested via the top electrode of the investigated GMR spin-valve structure (see Chapter 4). The field pulses are synchronized by means of an electronic delay t, which could be varied in steps of 10 or 20 ps, matching the overall time resolution[48]. For each image, the sample is thus excited and probed every 2 ns.

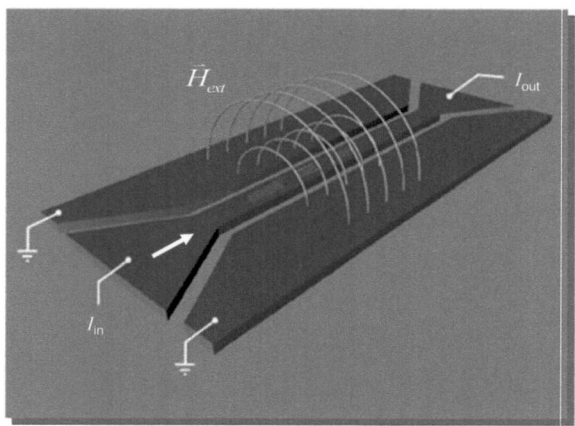

Fig. 10: Schematic principle of high frequency magnetic excitation of ferromagnetic platelets of various geometries (black) on a coplanar wave guide. The field lines depict the Oerstedt field $\vec{H}_{ext}$ generated by a current pulse I through the centre strip line (picture courtesy C.M. Schneider).

In the stroboscopic imaging mode the synchrotron radiation source is phase-locked with an electrical pulse generator. The current output from a commercial pulse generator yields a high-frequency electromagnetic field in a coplanar wave guide matched to 50 Ω impedance (see Fig. 11). In turn, a magnetic field H_{ext} (Oerstedt field) occurs around the central micro strip line. Above the central strip line the pulsed magnetic field H_{ext} field is sufficiently parallel to the surface to excite homogeneously the magnetic platelets deposited on top. The pulse generator used delivered pulse amplitudes of ~7 V with a full width at half maximum (FWHM) of ~100 ps (see Fig. 14a).

Snapshot images are obtained by varying the delay time between the onset of the rising edge of the magnetic field pulses and the photon pulses from the storage ring BESSY II. The in-plane component of the incident circularly polarized light points parallel to the exciting field thus revealing the dynamics of the magnetization component along the projection of the photon beam.

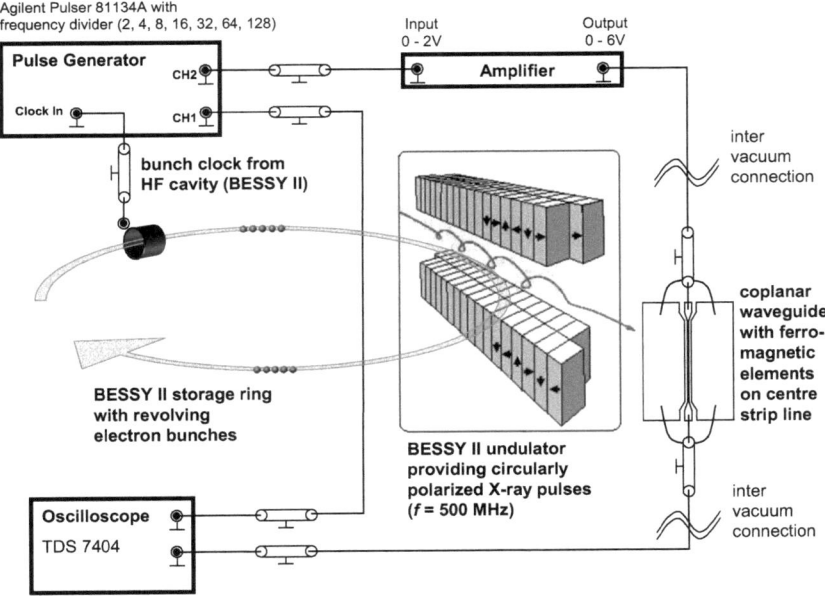

Fig. 11: Setup of the pump-probe experiment performed at BESSY II. The electrons revolving in the storage ring emit circularly polarized X-ray light when forced to a corkscrew trajectory by one of the undulators.

At the fundamental frequency of 500 MHz the fast field pulse shows considerable contributions of higher harmonics[24]. If the leading edge of the field pulse is sufficiently steep, then the magnetic response can be faster than the width of the magnetic field pulse. This means that the time resolution in this mode is essentially given by the width of the photon pulses from the synchrotron source and the electronic jitter of the experimental set-up. In the so-called low-α mode at BESSY II photon pulses of < 4 ps rms are provided via a special way of bunch compression[49]. The jitter measured by comparing synchrotron trigger and pulse generator output ranged between 8 and 14 ps. Together with the electronic jitter of our set-up it leads to a total time resolution of about 15 ps.

Fig. 12: Photography of the experimental setup of TR-XMCD-PEEM showing the vacuum chamber installed at beam line UE52-SGM at BESSY II. The beam line connects from the right hand side, the Focus IS-PEEM is flanged to the rear of the chamber.

3.2. Time-resolved XMCD-PEEM

The PEEM measures the spatial distribution of the X-ray absorption via the electron yield of secondary electrons. Fig. 12 shows a photo of the experimental setup at the UE52 beam line at BESSY II. The spatial resolution is of the order of 100 nm. When the energy of circularly polarized photons is tuned to the Ni–L_3 absorption edge ($h\nu = 853$ eV), the electron yield varies with the relative orientation of local magnetization M and polarization vector P. X-ray magnetic circular dichroism (XMCD) images are obtained from two images taken with opposite polarization calculating the asymmetry A_{XMCD} at each pixel, which is given by:

$$A_{\text{XMCD}} \propto \frac{I_{\text{right}} - I_{\text{leftt}}}{I_{\text{right}} + I_{\text{leftt}}} \propto \mathbf{P} \cdot \mathbf{M} \propto \cos\alpha \qquad (7)$$

with $I_{right/left}$ denoting the pixel's grey scale intensity for the right/left helicity and α being the angle between P and M. The observed contrast scales with the degree of circular polarization of the synchrotron radiation (P being projected into the plane of the platelet). The asymmetry equation (7) ensures that all contrast mechanisms other than magnetic cancel out, because they do not depend on photon helicity. To acquire an image, we typically integrate the signal for 30 s, thus integrating over 1.5×10^{10} pump-probe cycles. As the synchrotron radiation intensity declines between injections of electron bunches, the two raw images were first corrected for intensity variations before being fed into the asymmetry equation. The geometry of the experiment is sketched in Fig. 13.

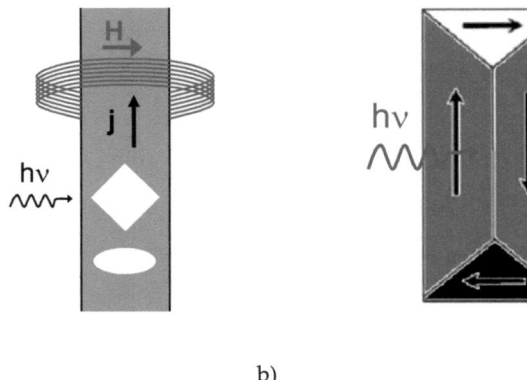

Fig. 13: a) Orientation of the centre micro strip line with ferromagnetic platelets in white to the incident circularly polarized light pulse. b) Example of the grey level appearance of the magnetic domains in a typical Landau-Lifschitz structure in a Py platelet in this geometry, determined by calculating A_{XMCD} for each pixel (eq. 7).

The rising edge of the exciting pulse is far shorter than its trailing edge, see Fig. 14a. From previous experiments, we know that the pulse shape on the micro stripline area observed in the microscope is narrower than measured with an oscilloscope behind the stripline outside the UHV system (incl. leads and contacts). Thus the rising edge of the field pulse ~ $H(t)$ can be expected to be significantly shorter than measured with the oscilloscope. Due to the fast rising edge, frequency components in the 10 GHz range are contained in the Fourier spectrum of the field pulse as shown in the Fourier decomposition, Fig. 14b.

The current pulse in Fig. 14a was reconstructed from the deformation of PEEM images of the particles being studied. The deformation is caused by the passage of the current pulse through the micro strip line. The generated Oerstedt field deforms the trajectories of secondary electrons emitted by the sample under UV irradiation.

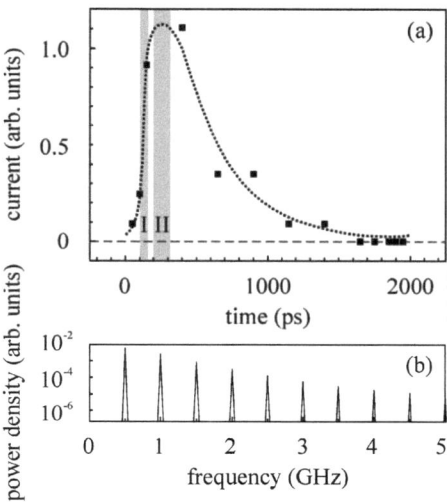

Fig. 14:(a) Temporal profile of the magnetic field pulse measured in situ by small image defocusing $\Delta f \neq 0$ (squares represent experimental values). The dashed line denotes a fit with a Fourier series. **(b)** Corresponding Fourier decomposition of the pulse profile[50].

The deformation is a measure of the field strength $H_{ext}(t)$. The black squares denote the experimental points. A high sensitivity of this method of measuring the pulse profile can be achieved by image defocusing $\Delta f \neq 0$. This is a special property of the PEEM optics. All following measurements of the domain patterns were performed at the best-focus condition $\Delta f = 0$. The Fourier spectrum (Fig. 14b) was determined by fitting the pulse profile (Fig. 14a) by a Fourier series

$$A(t) = \sum_n A_n \sin n\omega t, \qquad (8)$$

A_n denotes the amplitude of the Fourier component n. Obviously, the steep rising edge contains significant Fourier frequencies up to several gigahertz. Further details of the stroboscopic imaging approach performed in the photoemission electron microscope are described in references[13,48,51,52,53].

Chapter 4

Preparation of Micro Stripline Samples

For Permalloy platelets the magnetic reaction on the field pulse is expected to be dominated by the interplay of the domains of Landau-Lifschitz patterns like Fig. 2. In the focus of this thesis in contrast, are ferromagnetic samples with defined magnetic ground state, providing sufficient torque on the magnetization to restore homogeneous magnetization of the structure after the field pulse. A variety of coplanar wave guide samples have been prepared for this purpose besides standard samples with platelets of thicker Permalloy. Coplanar wave guide (CPW) samples with anti-parallel pinned synthetic GMR spin valves have been prepared as well as samples with platelets of single crystalline Co to create a uniaxial anisotropy pointing parallel to the strip line.

No.	Magnetic System / Orientation of CPW to plane of incident light	Geometry and Orientation of Platelet to the CPW	Scope of Experiment
1	GMR spin valve / vertical	Two rectangles with aspect ratio 2:1 with direction of pinning field parallel to centre strip line <u>Large platelet:</u> 10 × 15 µm² with long edge parallel to pinning direction. <u>Small platelet:</u> 5 × 10 µm² with long edge perpendicular to pinning field.	To create a magnetic ground state (pinning field by H_{exch}) parallel to centre stripe. To study the dynamics of the free layer independent from other FM layers in the stack by element specific imaging of Ni L_3 edge. Which role plays shape anisotropy?
2	GMR spin valve / vertical	Direction of pinning field parallel to centre strip line <u>Ellipse</u> (semi axes 6 µm × 3 µm): with horizontal orientation and <u>Square</u> (5 × 5 µm²): with 45° orientation on the CPW	Which role plays shape dependence and edge domains in the magnetic switching behaviour of GMR spin valves?
3	Ultrathin Py (d = 3 nm) / horizontal	Field induced anisotropy parallel to horizontal strip line <u>Ring</u> with inner diameter 6, outer 15 µm	To produce samples with Py layer as thin as possible to ensure full magnetic saturation of platelet by the field pulse. To investigate behaviour of ring in onion state.
4	Ultrathin Py (d = 3 nm) / horizontal	Field induced anisotropy parallel to horizontal strip line <u>Ellipse</u> (semi axes 6 µm × 12 µm): Orientation 45° to centre stripe.	To produce samples with Py layer as thin as possible to ensure full magnetic saturation of platelet by the field pulse. Avoidance of edge domains.
4	epitaxial Co / vertical (FIB preparation)	Two <u>rectangles</u> with aspect ratio 2:1 and uniaxial anisotropy along the centre stripe. (5 × 10 µm² and 10 × 5 µm²)	Co grown epitaxially with uniaxial anisotropy along centre stripe in order to achieve homogeneous magnetization.
5	Py (d = 10, 40 nm)	Rectangles featuring Néel walls	Observe high-frequency dynamics in the range of several GHz, search for spin wave propagation.

Table 1 Overview to the samples prepared in this thesis and their experimental scope.

Samples with ultra-thin Py deposited with field induced anisotropy were built with the same intention. Table 1 gives an overview to the samples investigated. Layouts of coplanar waveguides with horizontal and vertical orientation of the narrowed part of the micro stripline have been employed. For the spin valve samples, the vertical orientation of the coplanar waveguide was preferred in combination with a vertical pinning of the spin valve as will be discussed in detail in Fig. 22. This way the structure appears grey in it's ground state and contrast evolves as the platelet will be excited by the magnetic field. For the ultra-thin Py platelets on the other side, a horizontal coplanar waveguide geometry has been chosen in order to clearly verify the onion state of the Py ring without excitation and it's vanishing into grey as it becomes saturated.

4.1. Preparation of Micro Strip Line samples with Spin-Valves

Design Optimization

Sample preparation is crucial when it comes to real time observation of spin dynamics. Since the micro strip line is chosen to satisfy an impedance of 50 Ω in order to match with the high frequency electronics, the coplanar waveguide design - CPW geometry - is adequate (see Fig. 15). The advantages of this geometry are the little dispersion, resulting in little amplitude loss due to damping. Furthermore, coplanar waveguids can still be structured in one metallization layer, reducing the number of photo lithography steps during thin film wafer processing. The disadvantage is, however, that no pure transversal electromagnetic wave (TEM wave) exists in CPW geometry unlike in the screened triplet. At low frequencies ($\lambda < d\sqrt{\varepsilon_r}$), with d being the distance between the two areas on ground potential ($d = 2W + S$) and ε_r the dielectric constant of the substarte, the hybrid wave can be treated as a quasi-TEM wave since longitudinal components of H- and E-fields are small compared to transversal component and the TEM character remains preserved. At higher frequencies undesired higher modes occur (compare Fig. 17).

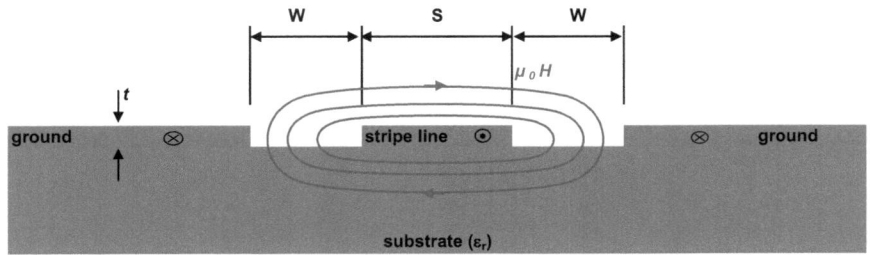

Fig. 15: Schematic cross section of the CPW micro strip line (not to scale) geometry used for sample preparation. Here, *t* is the metallization thickness, *S* the width of the centre strip line and *W* the width of the gap. The assumption of a dispersion free coplanar waveguide holds for *t* << *W*.

For $t \ll W$ the impedance Z_0^{cp} of the coplanar waveguide is a function of the ratio between the width of the current leading micro strip line and the gap to the adjacent areas on ground potential and also depends on the dielectric constant ε_r of the substrate.

Objective of sample simulation and preparation is to reduce damping and to increase the Oerstedt field acting on the ferromagnetic platelet on the centre stripe and thus the micro strip line's width was decreased down to $S = 20$ and 10 µm. According to Ampere's law, the integral along a closed line of the magnetic field equals the current passing through the field loop. For a one-dimensional micro stripline, with negligible height, Ampere's law reads as

$$I = \oint \vec{H} ds = H \cdot 2S \qquad (9).$$

Silicon wafers with high purity are necessary, since high dielectric constant is desirable ($\varepsilon_r = 11.7$) to suit strip line widths of $S = 10$ to 100 µm.

Two regimes exist for the impedance of a coplanar wave guide Z_0^{cp} as a function of the parameter k, defined as the ratio $k = S/(S + 2W)$, where S is the strip line width and W the gap width (see Fig. 16).

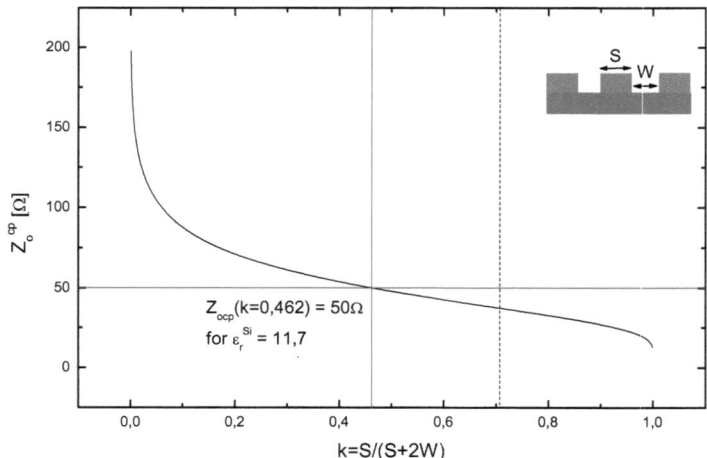

Fig. 16: Impedance as a function of parameter $k = S/(S+2W)$, where S is the strip line width and W the gap width). When designing the layout of samples on the wafer, S and W have to be chosen to fulfil $Z_0^{cp} = 50$ Ω, see horizontal line. The impedance in the two regimes below and above $k = 0.707$ (separated by the dashed line) is given by the formulas below.

$$Z_0^{cp} = \frac{30\Omega}{\sqrt{\varepsilon_{res}}} \ln\left[2\frac{(1+\sqrt{k'})}{(1-\sqrt{k'})}\right], \text{ for } k < 0.707 \qquad (10)$$

$$Z_0^{cp} = \frac{30\pi^2\Omega}{\sqrt{\varepsilon_{res}}} \ln\left[2\frac{(1+\sqrt{k})}{(1-\sqrt{k})}\right]^{-1}, \text{ for } k > 0.707, \qquad (11)$$

with $k' = \sqrt{1-k^2}$ and $\varepsilon_{res} = \frac{\varepsilon_r + 1}{2}$

Advanced software packages for dynamical electromagnetic simulation of microwave circuits (such as the Ansoft Designer[54]) have been utilized to confirm the waveguide design capabilities with respect to high bandwidth. Up to the frequency of $F = 20$ GHz theoretical damping and reflections could be reduced to the minimum, see Fig. 17. The oscillatory behaviour originates from higher order resonances of this specific coplanar waveguide design.

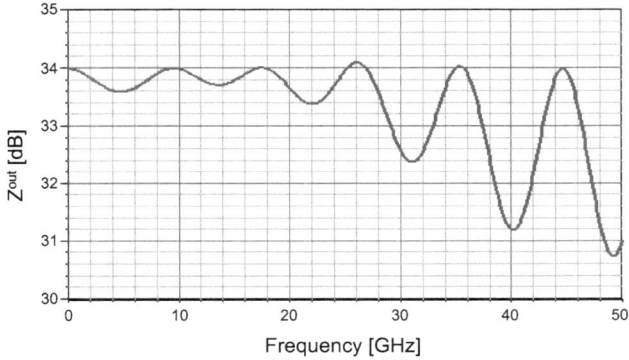

Fig. 17: Simulation results show marginal damping which is constant up to 20 GHz.

Device Fabrication

The deposition and photo lithography of metallization and insulator layers to structure CPW and the ferromagnetic platelets have been processed on a 5 inch silicon wafer technology in class 100 clean room environment at Sensitec GmbH, Mainz. In a first step (see Fig. 18, step1) 300 nm Al_2O_3, a proven insulator in thin film technology with high dielectric strength (16.7 kV/mm), is deposited as full film onto the silicon wafer substrate by cathode magnetron sputtering on a Unaxis Corona deposition system.

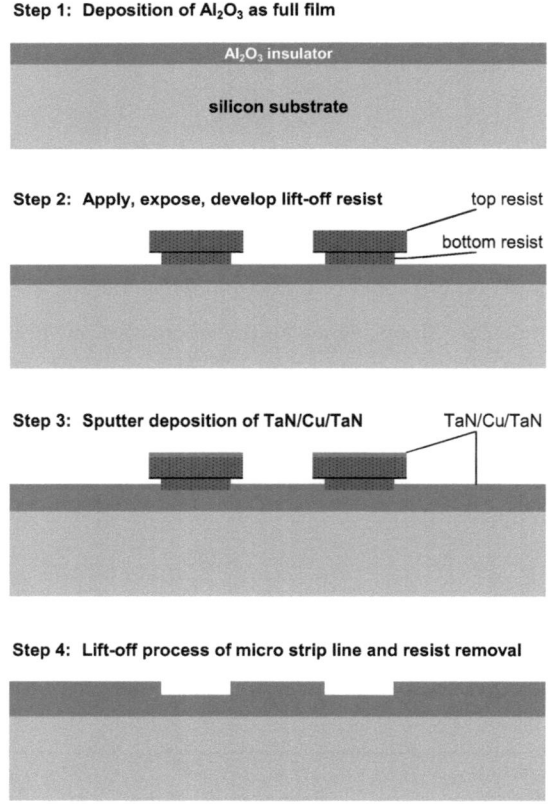

Fig. 18: Schematic sequence of process steps on wafer level for preparation of the coplanar wave guide: Step1 shows the full film deposition of the first Al_2O_3 insulation layer. In steps 2 to 4 the structuring of the coplanar wave guide with lift-off technique is depicted.

The deposition and structuring of the coplanar wave guide lead layer takes place in steps 2 to 4. Here, reactively sputtered TaN serves as a diffusion barrier. Structuring the coplanar wave guide layer has been done by using lift-off technique with a two-layer positive photo resist. This technique is commonly applied in photo lithography when materials have low ion etching rates, like Ta. The photo resist will be applied first, exposed and developed to leave a negative lift-off resist system of the desired coplanar wave guide structure on the wafer. Exposure of the two-layer photo resist utilizes the i-line, i.e. $\lambda = 365$ nm, of a mercury light source in Canon 5i+ lithography tool. This way, critical dimensions of > 0.78 µm can be realized. The two layer photo resist is used because an undercut (or recession) of approx. 150 nm is favourable for the lift-off process, since openings in metallization at the foot of the resist system allow access to the wet chemical solvent N-Methyl-2-pyrrolidinone (NMP) (see Fig. 18, step 2). The recession emerges during development of the dual layer resist system because the bottom resist has been chosen with a slightly higher solubility than the top resist. Hereafter the coplanar wave guide is deposited full film as a tri-layer consisting of TaN/Cu/TaN by using state of the art cathode RF sputtering technique (Unaxis Emerald Z660) (see Fig. 19, step 3). The lower TaN layer with a thickness of 20 nm serves as an adhesive promotor to the underlying Al_2O_3. The Cu layer is 250 nm thick. The upper TaN layer is deposited with an increased thickness of 100 nm to serve as an etch stop for the subsequent ion milling process (see Fig. 19) and as a diffusion barrier for Cu simultaneously. By finally detaching the complete lift-off resist system with the metallization on top by application of NMP, only the desired coplanar wave guide structure remains (see Fig. 18, step 4).

Steps 5 to 9 in Fig. 19 depict the deposition and structuring of the ferromagnetic layer. In case of the GMR spin valves this takes place in a state of the art PVD (Plasma Vapour Deposition) - IBD (Ion Beam Deposition) hybrid cluster tool (Connexion by CVC, Inc.). The hybrid cluster tool combines three vacuum deposition chambers with an in-situ wafer handling system as shown in Fig. 20.

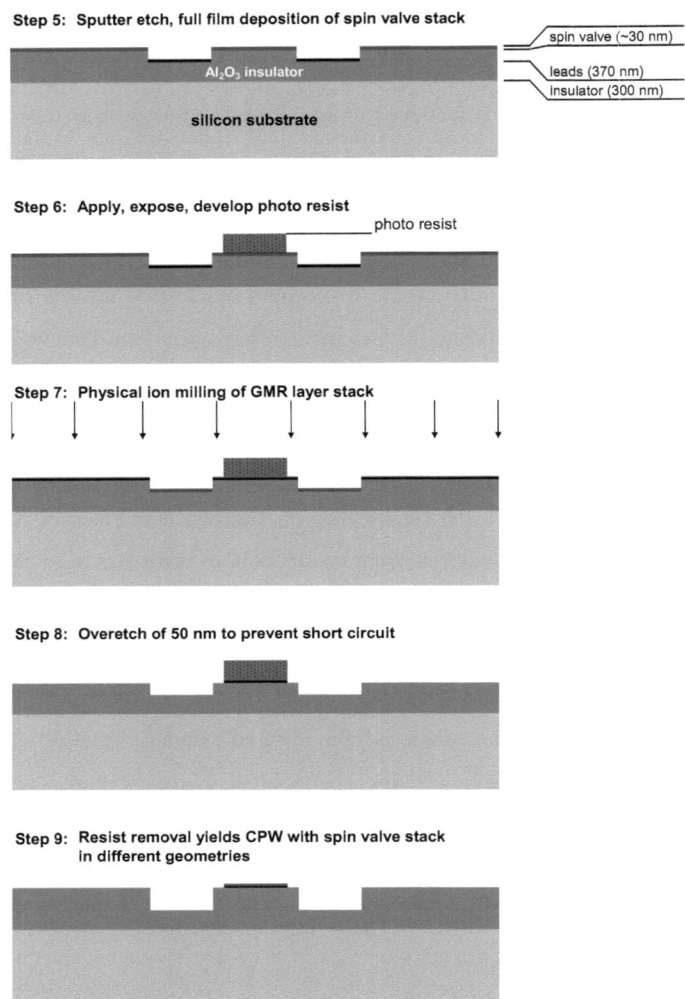

Fig. 19: Schematic sequence of process steps on wafer level for preparation of the magnetic elements: Full film deposition of the GMR spin valve stack (step 5) and structuring into desired geometries by ion milling with a positive photo resist.

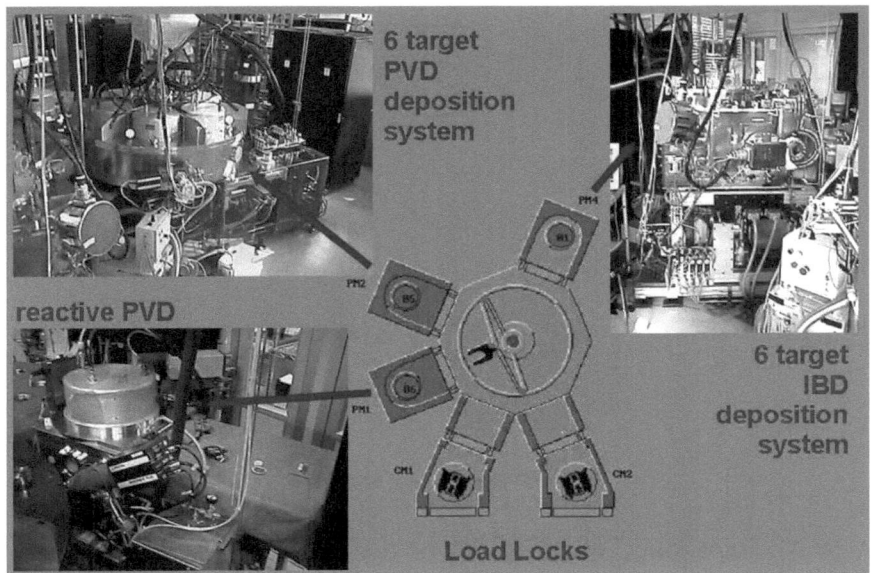

Fig. 20: Cluster tool combining three vacuum deposition chambers with an in-situ wafer handling system. The PVD and IBD chambers comprise 6 different metal targets each (magnetic and non-magnetic). Al_2O_3 can be deposited by the additional reactive PVD chamber.

The physical vapour deposition (PVD) chamber comprises six different metal targets, magnetic and non-magnetic. The integrated ion beam deposition system (IBDS) allows deposition from two different metal alloy targets and is mainly used for seed layer deposition. Established ferromagnetic alloy metals are $Ni_{89}Fe_{19}$, $Ni_{55}Fe_{45}$ or $Co_{90}Fe_{10}$. Pure non-magnetic metal spacers can be deposited from copper or ruthenium targets. A third PVD module with a pure aluminium flanged to the system allows Al_2O_3 deposition in a reactive process.

The antiparallel pinned synthetic GMR spin valve was deposited full film onto the structured CPW (see Fig. 19, step 5). The complex layer sequence is Al_2O_3 / PtMn / CoFe / Ru / CoFe / Cu / CoFe / NiFe / Ta (see Fig. 21). The overall thickness of the complete stack does not measure more than 300 nm. The thickness uniformity across the wafer of each deposited layer is vital to the magnetic performance of the GMR spin valve stack and is below 3%. The thickness of the Cu spacer is tuned to the second antiferromagnetic coupling maximum (AFCM).

a)

b)

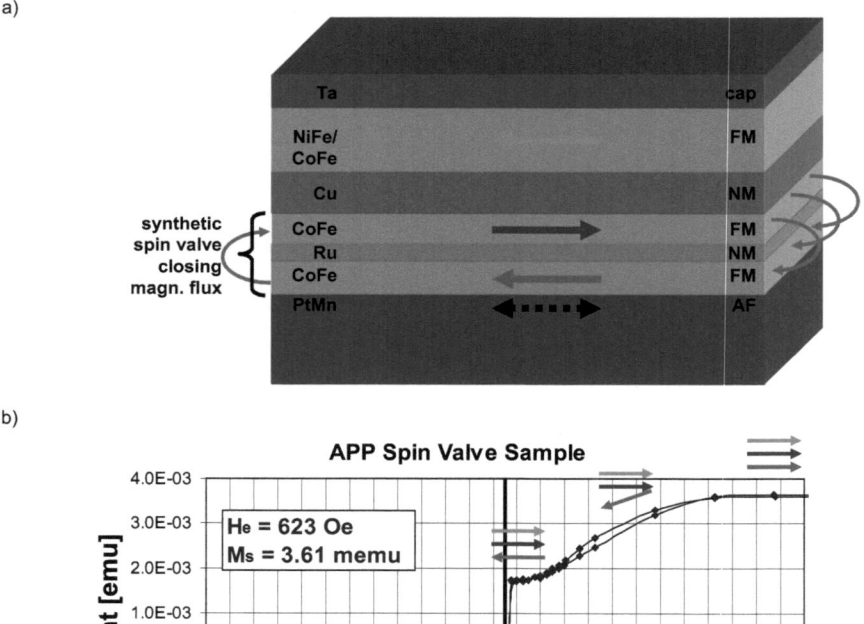

Fig. 21: a) Schematic layer stack of the antiparallel pinned synthetic spin valve studied in this thesis. The coloured arrows indicate the magnetization direction of the two hard magnetic CoFe layers of the synthetic spin valve (red and blue) and soft magnetic free layer (green) in the ground state. b) Typical spin valve magnetization curve measured from -5000 to 5000 Oerstedt with vibrating sample magnetometer (VSM). The arrows show the switching and rotation processes of all three ferromagnetic layers. The method is described elsewhere[55].

Especially the spacer thicknesses (Ru and Cu) require these constraints to achieve a reproducible GMR effect of typically 15% $\Delta R/R$ as shown in Fig. 9.

In a more detailed description of step 6 in Fig. 19 it splits up into a pre bake process step of substrate, the application of the positive photo resist, the exposure with UV light at the *i* - line through the reticle with the magnetic structures, the development of the resist followed by several inspections of the development process and the measurement of the accuracy of the alignment to the preceding layer. The mask design provides alignment structures that assure realignment of subsequent layers with accuracy better than 150 nm. For PEEM investigations, this accuracy is adequate, since any significant displacement of the magnetic platelets from the middle of the strip line causes an inhomogeneous magnetic field at the platelet which in turn leads to undesired distortions in the PEEM image.

Removing the spin valve stacks from the areas not covered by photo resist takes place in a dry physical process, since the uniformity of ion milling rate across the wafer is superior to wet etching processes. The process of removing material by bombardment of the wafer with a broad beam of Ar^+ ions is often referred to as physical ion milling (see Fig. 19, steps 7,8). To assure that no conducting residues remain on wafer the ion milling is continued into the underlying Al_2O_3 layer until a depth of 50 nm has been reached.

In step 9 of Fig. 19 the photo resist will finally be removed in the solvent NMP. After finalization of the wafer in the wafer front end, the back end processes comprise mounting the wafer on dicing foil and die separation. Wafer backside thinning has been abandoned, since a chip height of the basic substrate thickness of 625 µm is favourable to the experiment. The distance of sample wiring to the extractor lens of the PEEM can be increased this way.

As stated above, the samples studied stand as examples of complex magnetic layer stacks and represent advanced spin-valve structures, designed to optimize the GMR effect. They have been successfully implemented into commercial devices by Sensitec GmbH. The sequence of the layer stack as shown in Fig. 21a) depicts the antiparallel alignment of the magnetization direction of the three ferromagnets in ground state. Here, the two hard magnetic CoFe layers form of the synthetic spin valve (red and blue) closing the magnetic flux. The magnetic pinning of the lower CoFe layer to the adjacent antiferromagnet PtMn emerges when annealing the system above the antiferromagnets's Néel temperature of 275°C for several hours and simultaneously applying a magnetic field above 1 T. The effect evolves with the formation of an interdiffusion layer at the boundary where local magnetic moments freeze due to surface roughness.

The soft magnetic free layer consists of the bilayer CoFe/NiFe in order to combine to important features of the spin valve. By deposition of CoFe first, the exchange coupling of to the underlying CoFe is higher than to a NiFe directly. The NiFe in turn will be deposited on top of the CoFe due to it's soft magnetic properties. The resulting transfer curve of such a spin valve reveals that little field is required to flip the free layer magnetization and drastically change the resistance. A complete transfer curve measured from -5000 to 5000 Oerstedt with vibrating sample magnetometer (VSM) is shown in Fig. 21b. The arrows show the switching process of all three ferromagnetic layers in the layer stack.

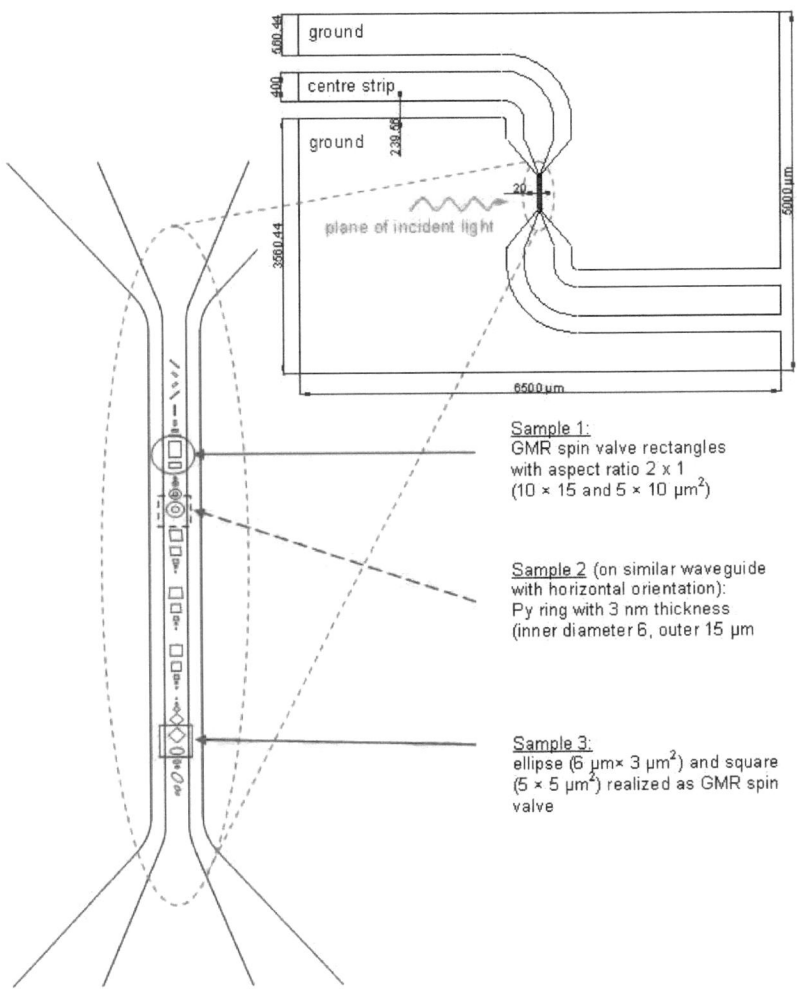

Fig. 22: Coplanar waveguide with vertical orientation of the bottle neck. In this area, the centre strip line is narrowed down to 20 µm and equipped with ferromagnetic microstructures with various geometrical shapes. The structures (samples 1 and 3, indicated by solid arrows) on this particular waveguide design, have been deposited as multilayer stacks for GMR spin valves (see Fig. 23). The geometries chosen for investigation are rectangles in two different dimensions and orientations to the Oerstedt field but with fixed aspect ratio (sample 1) and a pair of a diamond and an ellipses (sample 3). On a similar waveguide with horizontal orientation of the bottle neck ultra-thin Permalloy platelets have been deposited. The investigation on this sample focused on the ring-shaped platelet (sample 2, indicated by dashed arrow) and the ellipse under 45° orientation.

The magnetically soft CoFe/NiFe free layer is separated from the CoFe hard layer by an ultrathin Cu interlayer providing an antiferromagnetic coupling field of 0.6 mT, as derived from the easy axis loop ($H \parallel y$). The hard axis loop ($H \parallel x$) reveals a nearly reversible magnetization rotation. From the initial slope of the hard axis loop, we deduce a total anisotropy field of 1.5 mT. The difference might be ascribed to a uniaxial anisotropy due to the field applied during the sample preparation.

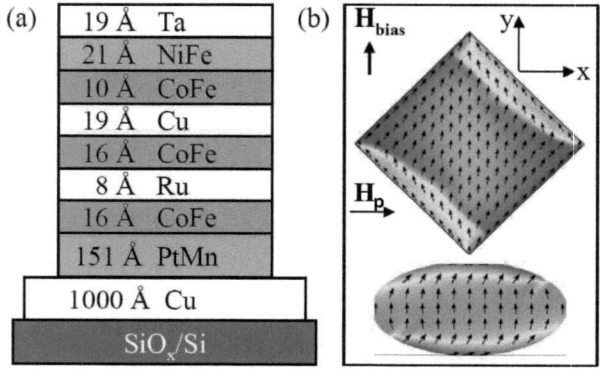

Fig. 23: Sketch of the sample geometry as a cross section (a) and top view (b). Detailed sequence of synthetic spin valve stack. Snapshot of the micromagnetic simulation at 1100 ps indicates edge domains resulting in an "S" shaped domain configuration of the free layer. H_{bias} and H_p denote the exchange bias and the pulse field directions, respectively[56].

The detailed information about the thicknesses of the layers in the spin valve stack can be obtained from Fig. 23. Micromagnetic simulations of the free layer already indicate the existence of edge domains and their dependence on the geometry of the platelet.

All samples were mounted on UHV-exchangeable high-frequency compatible holders in the way shown in Fig. 24. Standard micro sample holders were modified to host two high bandwidth UHV compatible coaxial connectors. The stripline samples were mounted on a socket, adapting the distance to the extractor lens of the electron microscope. This way imaging at high extractor voltages was possible without arc discharges. With increasing extractor voltage, the resolution increases, the electron yield rises and emitted electrons are less sensitive to distortions on their trajectory. This leads to a high signal to noise ratio in the image which is beneficial particularly when working at high magnification.

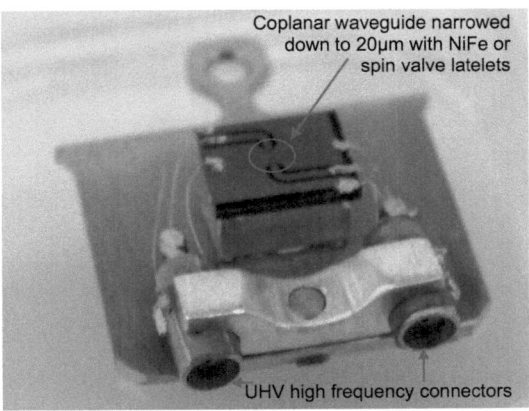

Fig. 24: UHV-exchangeable sample holder with the high frequency connectors and the sample chip comprising the coplanar wave guide and the ferromagnetic platelets on the centre strip line narrowed down to 20 μm (seeFig. 22).

4.2. Preparation of Single Crystalline Co Platelets

The single crystalline Mo(110) striplines were produced in the following way. In the first step, a Mo(110) epitaxial film (200 nm) was deposited onto Al_2O_3 in UHV by electron-beam evaporation. During molybdenum evaporation, the temperature of the substrate was kept constant at 1000 K. In the second step, the strip-line (coplanar waveguide with a width of 100 µm of the central lead) and contacts were patterned by standard photolithography. The sample was again transferred into the UHV chamber and in the third step, the Mo strip-line was cleaned by repeated cycles of heating in oxygen atmosphere up to 1300 K, followed by flashing at 1800 K in UHV until the Auger electron spectroscopy (AES) signals of C and O originating from the Mo surface was below the detection limit. On this sample a Co layer (20 ML) and a Au capping layer (5 ML) was deposited by molecular beam epitaxy (see Fig. 25a, b).

Epitaxial growth of the Co layer was controlled by low energy electron diffraction (LEED). The Co/Au layer was kept thin enough to avoid electrical shorting of the strip line.

The structure of the Mo(110) strip-line and the Co layer were investigated using low energy electron diffraction (LEED). Atomic force microscopy (AFM) and scanning tunnelling microscopy (STM) were used for the characterization of the morphology and chemical purity of the samples was verified by Auger electron spectroscopy (AES). The magnetic properties of Co films were investigated ex-situ using magneto-optical Kerr magnetometry (MOKE).

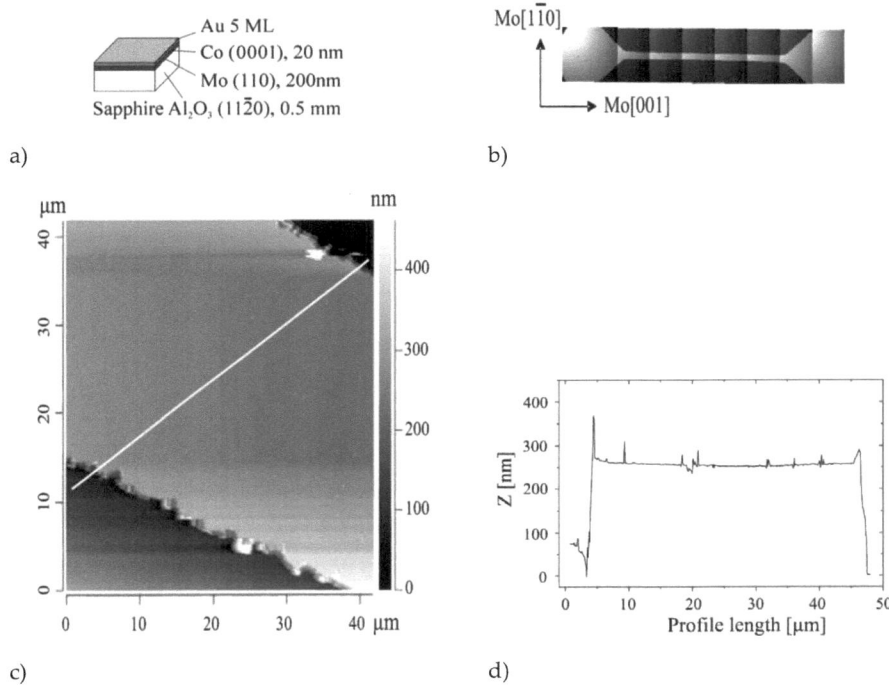

Fig. 25: Sample preparation for time-resolved XMCD-PEEM of epitaxial Co platelets on a single-crystalline Mo($1\bar{1}0$) strip line on Al$_2$O$_3$($11\bar{2}0$). a) Schematic layer sequence on sapphire substrate: Au/Co/Mo/Al$_2$O$_3$ d) Image of the narrow part of Mo($1\bar{1}0$) strip-line of a width of 38 μm and length of 1mm on the sapphire substrate, image taken using optical microscopy after the photolithography procedure is finished (field of view 1.8 × 0.3 mm^2). c) AFM image (42 × 42 μm^2) of the strip-line; the white line denotes the scan profile shown in d). Spikes observed along the scan are due to the dust particles[53].

We found that the surface of the Mo strip-line is smooth in a large scan area (Fig. 25c). Small grains originating from residues of the photo resist layer and dust particles are observed. They were removed during the cleaning of the substrate in vacuum. The STM measurements performed in situ after the third step of the preparation procedure, show that the molybdenum surface consists of monoatomic terraces of widths varying between 10 and 200 nm (Fig. 26a), comparable to the widths of terraces observed on single crystal Mo surfaces[57,58]. The LEED pattern (b) indicates the bcc (110) structure of the Mo strip-line surface. Sharp diffraction spots visible in the LEED pattern confirm a well ordered surface structure of the strip line. We conclude that the photolithography and etching procedure did not destroy the morphology and the structure of the molybdenum buffer layer on sapphire.

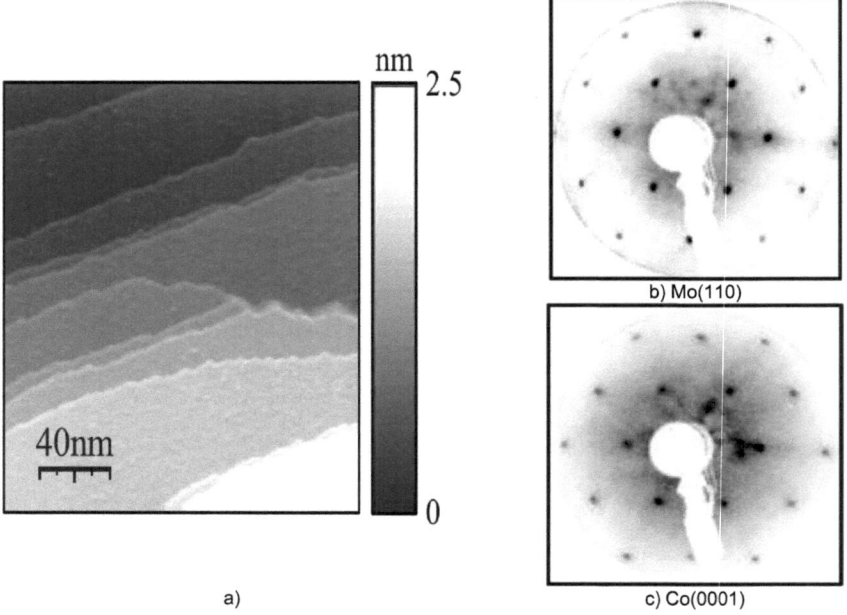

Fig. 26: a) STM image (200 × 200 nm^2) of the Mo(110) strip-line surface obtained after the last step of the preparation procedure in UHV, before deposition of the cobalt layer. b) LEED pattern originating from the Mo(110) strip-line surface obtained at E_p = 218 eV, and c) LEED pattern obtained for the Co(0001) film (thickness 20 nm) deposited on the Mo strip-line surface at E_p = 254 eV [53].

The MOKE measurements were performed at different positions of the sample. Magnetic hysteresis loops obtained for the Co films deposited on the bare part of the Al$_2$O$_3$ substrate differ from the magnetization loops measured for cobalt films prepared on the Mo strip-line, demonstrating that the Au/Co/Al$_2$O$_3$ films do not reveal any magnetic anisotropy. It can be expected that the two-fold symmetry of the bcc single crystalline Mo strip-line substrate induces a uniaxial in-plane magnetic anisotropy in the cobalt film, as known from the results of the magnetic measurements for Co films, prepared on bcc (110) single crystals[59]. We observed a uniaxial anisotropy as illustrated by the magnetization measured using the magneto-optical Kerr rotation (MOKE) $\theta_K(H)$ (Fig. 27a and b). The magnetic easy axis of the Co film lies in the Co[1$\bar{1}$00] direction parallel to Mo[1$\bar{1}$0], while the hard axis is directed along the Mo[001] direction.

It has to be noted that the hard axis magnetization loop (Fig. 27b) deviates considerably from the reversible behaviour, as one would expect from the minimization of the free enthalpy density[60]. A similar, non-reversible loop was observed by Osgood et al.[61,62], and explained by a non-linear contribution to the longitudinal Kerr rotation $\theta_K(H)$, resulting from the product of the longitudinal and transversal magnetization components M_l and M_t:

$$\theta_K(H) = aM_l(H) + bM_l(H)\,M_t(H) \qquad (12)$$

The final structuring of the platelets on the central lead was performed by focussed ion beam etching (FIB). I also narrowed the central lead on a length of 1 mm to a width of 20 µm in order to increase the Oerstedt field according to eq. 9. In analogy to a scanning electron microscope SEM, a focused ion beam is scanned across the sample surface removing material by physical ion bombardment. Beam downwards, the ion optical system consists of a primary electromagnetic lens, a lens aperture, beam blanking plates for x/y-deflection an octupole for stigmation and a final focussing lens. Today, FIB is most commonly employed for a large variety of applications. Primarily, the technique is indispensible for defect analysis and process control in wafer level manufacturing due to it's capability to prepare micron-sized cross sections of thin films layer stacks. Other applications are device modification of CMOS chips where electrical circuits are manipulated by cutting away connections and drawing new ones by e.g. Tungsten deposition, TEM lamella preparation[63], micro structuring (e.g. trimming the magnetic joke of a hard disk write head) or magnetic patterning of exchange coupled multilayers[64,65].

A liquid metal ion source (LMIS) provides a DC beam of Ga$^+$ ions (see Fig. 28b). Gallium is an established metal for ion beam production since its melting point lies only 5 K above room temperature (T_{melt} = 29.76°C).

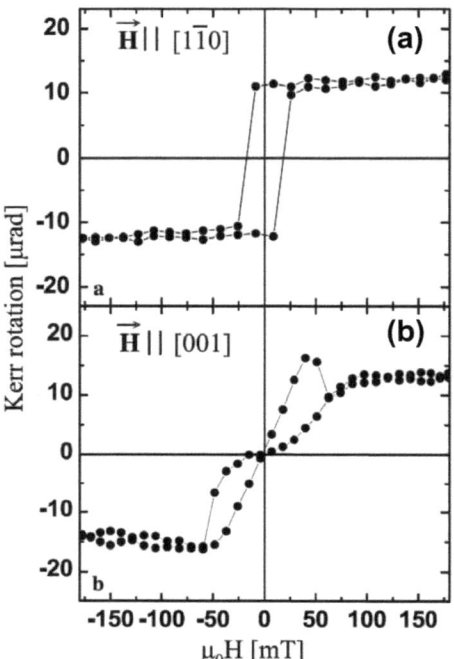

Fig. 27: a) Easy axis magnetization loop measured for a 15 ML Co film deposited on the Mo(110) strip-line. The MOKE measurement was performed in the longitudinal geometry and the magnetic field was applied along the Mo[1$\bar{1}$0] ∥ Co[1$\bar{1}$00] direction. b) Hard axis magnetization loop for the same Co layer as in a), with the external magnetic field applied along the Mo[001] ∥ Co[11$\bar{2}$0] direction[53].

For all applications mentioned, the stability of the ion beam is of high importance. At the tip of the heated Ga reservoir a drop of liquid Ga forms from which ions are extracted by a ring shaped extractor (typically on V_{extr} = 1 to 3 kV against LMIS). The high stability performance is achieved by a permanent feed back loop adjusting the suppressor with opposite polarity to a constant Ga⁺ ion current. This way the beam current instability is less than 5% of total beam current per hour[66]. The primary Ga⁺ ion beam in this application has a DC current of I = 2.2 µA. A number of current apertures ranging from 1 µA to 20 nA allow tuning the beam current to the desired etch rate.

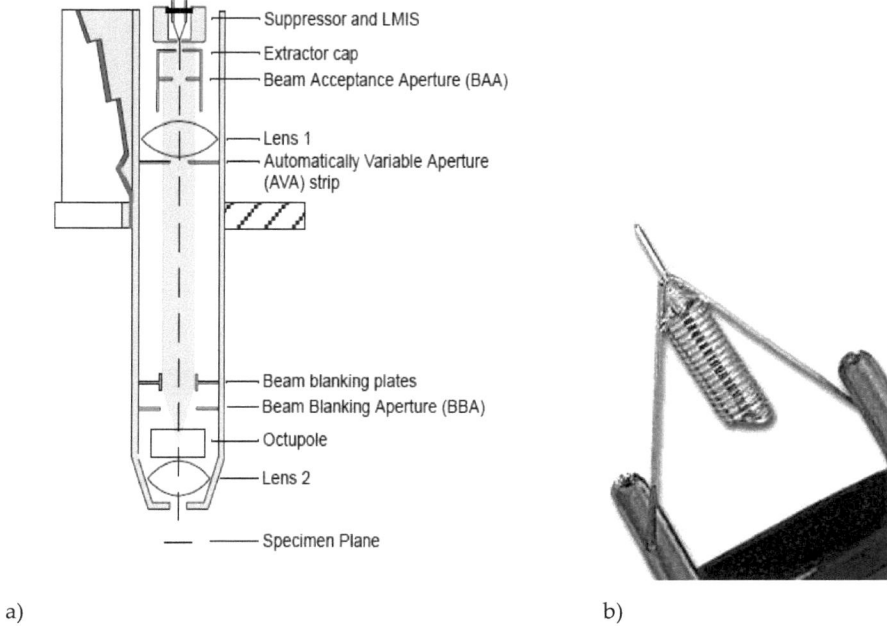

Fig. 28: a) Generation of a focused Ga$^+$ ion beam from a liquid metal ion source. Just as in scanning electron microscopy, the beam can be focused, scanned across the sample and blanked off the sample. **b)** Liquid metal ion source (LMIS). The heated reservoir contains liquid Gallium. The tip of source is that way constantly wetted with a hanging drop of liquid Ga of which the ions are extracted[67].

The tool employed is a FEI Altura 865 dual beam workstation, integrating an SEM column with a field emission gun (FEG-SEM) with the focused ion beam column. By SEM imaging of the surface it can be assured that no Mo or Co residues remain after ion milling. This way the etch rates of Mo and Co did not have to be determined, beforehand. The resolution of the focused ion beam is $d = 7$ nm at $V_{acc} = 30$ kV and the stage accuracy is better than 1.5 µm[66]. Fig. 29a depicts SEM micrographs of the area which has been additionally narrowed with FIB to increase the Oerstedt field (red dashed circle). The enlargements of this region show the 20 µm bottle neck in the micro stripline of originally 100 µm (Fig. 29b-c). In a further enlargement (Fig. 29d) the remaining Co platelets appear bright above a dark background.

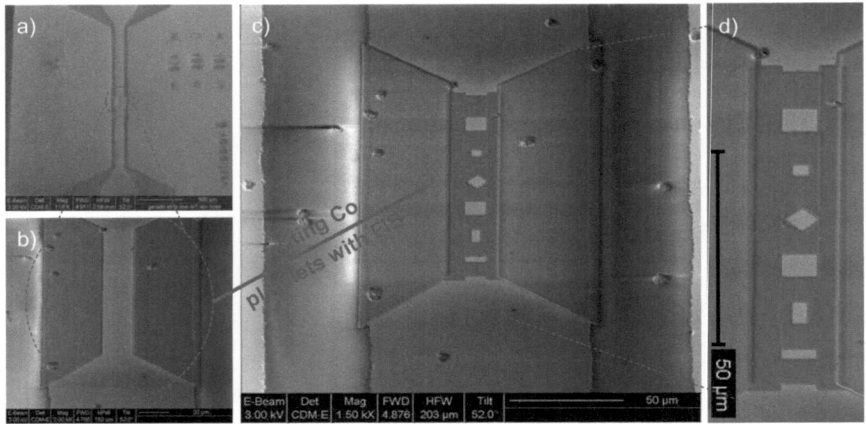

Fig. 29: Preparation of single crystalline Co(0001)Mo(110) strip line with focused ion beam technique. Images taken by scanning electron microscopy. a) Micro strip line narrowed down with FIB to 20 μm (dashed circle) at 118 x magnification. b) Enlargement of the bottle neck at 2000 x magnification. c) Region of centre strip line with Co platelets remaining after Ga$^+$ ion bombardment (magnification 1500). d) Further enlargement of single crystalline Co platelets structured by FIB.

4.3. Ultra-thin Permalloy Rings and Ellipses

Ion Beam Deposition (IBD) has been chosen as deposition method for soft magnetic monolayer films for the numerous advantages of the technique. Unlike conventional cathode sputtering, in ion beam deposition neither target nor substrate are exposed directly to the plasma. A broad beam of noble gas ions is extracted from the Kaufmann ion source[68] and directed to the metal target. The distances between ion source and target and between target and substrate are greater than for cathode sputtering and thus better chamber vacuum can be achieved at target and substrate. This way the temperature of the substrate is not affected by the plasma. Whereas noble gas ions are embedded into the metal film during conventional sputtering, IBDS deposited metal films stand out due to their high purity and the advanced reproduction of the alloy composition on the wafer with respect to the target. Another advantage of IBD systems is the possibility to employ ion assist sources, which are directed to the substrate to control film growth. The energy of the noble gas ion of the assist source is typically chosen below the threshold of sputtering. This way the impact of the ions is small enough not to dissipate ions off the growing film, but high enough to increase the mobility of the atoms in the growing film. For ferromagnetic metals, this typically leads to significantly reduced surface roughness and a higher magnetoresistive effect of IBDS ferromagnets than cathode sputtered ones

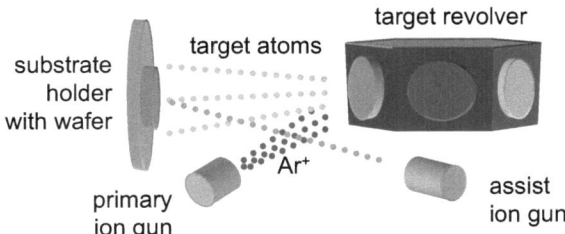

Fig. 30: Schematic view of an ion beam deposition system with Kaufmann type ion source. The Ar$^+$ ions generated by the primary gun dissociate metal atoms from the target, which in turn contribute to thin film growth on the wafer substrate.

Finally, commercial IBD systems stand out due to the excellent film thickness distribution of less than 2-3 %.

We were aiming at Py thicknesses as thin as possible in order to create platelets which would show full magnetic saturation during the field pulse. The ferromagnetic Permalloy structures deposited by IBDS have a thickness of $d = 3$ nm capped with 1.5 nm of Ta as protective coating, which will be removed in vacuum by smooth sputter etching prior to the synchrotron experiment.

4.4. Thicker Permalloy Platelets

The standard samples with thicker Permalloy platelets have been prepared by the group of Prof. C.M. Schneider with the objective to investigate Landau-Lifschitz patterns. Here, the micro strip line of 50 µm width was prepared by means of photolithography and subsequent wet chemical etching of a 250 nm thick Cu film on a SiO_x/Si substrate. A bilayer (40 nm $Ni_{81}Fe_{19}$, 2 nm Cu) was deposited on top of this micro strip line by DC magnetron sputtering and was micro structured by optical lithography and ion milling. The Cu cap layer served as a protection against oxidation and was removed by mild ion etching after introducing the sample into the photoemission electron microscope. The measurement were performed on two rectangular elements of 40×20 µm^2 and 40×10 µm^2 in size, i.e. aspect ratio of 2:1 and 4:1, respectively.

Chapter 5

Results

5.1. Magnetization Dynamics in Microscopic Spin Valves

We first recall the behaviour for a single *layer of a soft magnetic material*. Standing magnetic eigenmodes have been studied by Krasyuk[69] in a rectangular permalloy ($Ni_{80}Fe_{20}$) platelet of 16 x 32 µm² size and 10 nm thickness placed on a coplanar Cu waveguide, XMCD–PEEM snapshot images are shown in Fig. 31. Because of the incidence of the X-rays at 65° off normal, we are sensitive to the in-plane magnetization component along the short side of the platelet (*x*-direction). In the field-free state the equilibrium magnetization is a symmetric flux-closure domain pattern, comprising two equally sized domains separated by an 180° domain wall and two end domains as shown in Fig. 31e (dashed lines). In Fig. 31a, taken in the dynamic mode, the domains at the top and bottom of the platelet are oriented parallel and antiparallel to the circular photon polarization P_{circ} (along $h\nu$) and thus appear black and white.

The two domains oriented perpendicular to P_{circ} both appear grey and a 180°-Néel wall along the y-axis separates these domains with *M* upward (left) and downward (right). In the Néel wall, *M* is oriented to the left; thus the wall appears black. The external field H_{AC} is applied along the short side of the platelet, thus exciting a forced precessional mode of the magnetization with a frequency of 1 GHz (the first overtone of H_{AC} that is phase locked to the storage ring running at 500 MHz). The time evolution of the magnetization *M* in the platelet was studied for three different amplitudes of the magnetic field. Fig. 31b–d shows three snapshots taken at the instant of maximum magnetic response for field amplitudes I, II and III (0.15, 0.2 and 0.25 mT). The intensity in the two large domains varies periodically, indicating the predominant excitation mode of a precession of the magnetization around the effective field axis directed parallel to the long side of the platelet (*y*-axis). The system resembles a driven oscillator, and we observe its dynamical answer to the periodic excitation. The behaviour is reproduced by dynamic simulations in very good agreement.

In the next section we will see that the experimental results of a magnetically pinned multilayer stack differ substantially from the soft magnetic single permalloy layer.

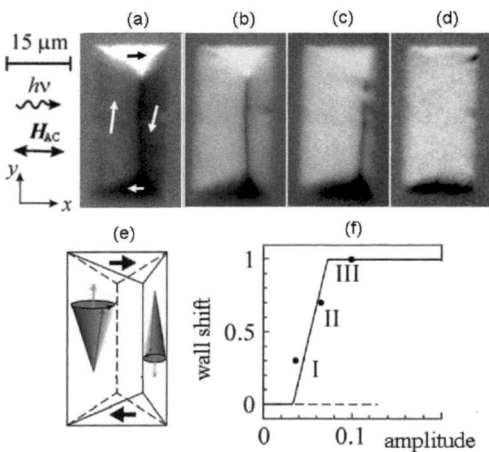

Fig. 31: Magnetic response of the x-component of the magnetization (bright areas are magnetized to the right, dark areas to the left) in a permalloy platelet of 16×32 µm^2 size and 10 nm thickness for three different field amplitudes I (0.15 mT), II (0.2 mT) and III (0.25 mT). (a) XMCD-PEEM snapshot of the domain pattern in dynamic mode at excitation amplitude I; arrows denote the local magnetization direction. (b-d) Snapshots of magnetic domain patterns at maximum magnetic response excited with increasing amplitudes (I, II and III for b, c, and d, respectively). (e) Sketch of the magnetic domain pattern in the field-free state (dashed) and when excited with the AC magnetic field (full lines). (f) Comparison of the experimentally observed mean shift of the central domain wall (dots) with an analytical model (full curve)[70].

Fig. 32 shows a similar measurement as shown in Fig. 31 but for an *antiparallel pinned synthetic spin valve* as described in Chapter 4. In the ground state, the exchange bias field forces the microscopic spin valve elements into an almost uniform magnetization state (weak contrast in Fig. 32, 0 ps). Only in the vicinity of the edges the magnetization turns parallel to the boundaries, comprising a positive (negative) value of $M_x(t)$ and avoiding stray field energy.

The magnetic field pulse of amplitude $\mu_0 H_p$ = 1 mT rotates the magnetization $M(r,t)$ into the direction of the external field. After the pulse has passed, $M(r,t)$ rotates through the equilibrium position into the opposite direction and finally back to its initial direction.

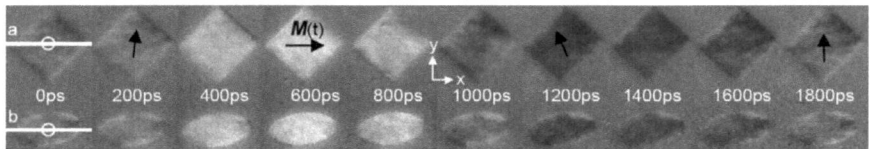

Fig. 32: Sequence of snapshots of the XMCD contrast of a quadratic (5 ×5 µm²) and elliptical (6 µm×3 µm) spin-valve element acquired simultaneously at the indicated time delays. The external pulse field (amplitude $\mu_0 H_p$ = 1 mT with time dependence according to Fig. 23 is applied along the horizontal x- axis. The easy magnetization direction points along the perpendicular y- axis. The grey level indicates the magnetization component along the x-axis. For some delay times, the magnetization vector in the centre of the square particle is indicated by arrows[56].

In order to test the homogeneity of the precession across the structure, we analyze line profiles taken along the diagonal of the square and the long axis of the ellipse, see white lines in Fig. 32.

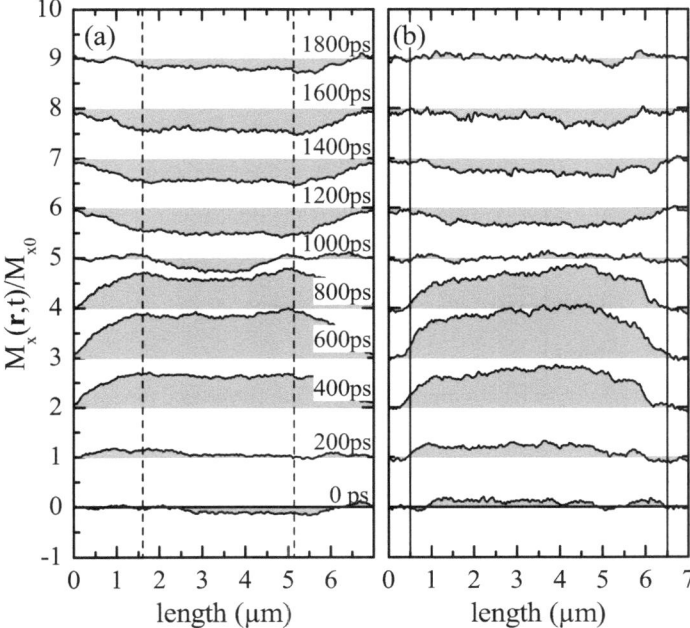

Fig. 33: Snapshots of the profiles of the magnetization component M_x/M along the lines (a) and (b) indicated in Fig. 32 at the designated time delays. The magnetization component $M_x(r)$ is roughly calibrated by the maximum XMCD value M_{x0} measured during the sequence of images. The dashed vertical lines in (a) denote extreme positions, see text. The solid lines in (b) denote the boundaries of the elliptical particle. For clarity, the profiles are vertically shifted by unity[56].

These profiles are shown in Fig. 33a and b for the square and ellipse, respectively. The profiles reveal that $M(r,t)$ is not phase coherent in the case of the square. After an almost homogeneous initial rotation toward the field direction (x), $M_x(t)$ starts to decrease at 600 ps in the central part faster than in the two areas close to the corners. This incoherent rotation leads to the wave like pattern at 1000 ps comprising two nodes along the (x) diagonal. Then, with increasing delay time, $M_x(t)$ increases faster in the centre, resulting in the two separated minima (occurring at the same position as the maxima observed at 400–1000 ps). Finally, $M_x(t)$ again reaches a rather homogeneous value across the square at 1800 ps. In contrast, similar profiles across the ellipse shown in Fig. 33b reveal an almost coherent rotation of $M(r,t)$ indicated by an almost constant value of $M_x(t)$ for fixed delay times.

As snapshot images have been taken all 20 ns, the magnetic response could be traced with high precision as shown in Fig. 34. The circles in Fig. 34a represent the profile of the field pulse. The response of the component M_x/M is shown in Fig. 34b comparing the time dependence of $M_{x,m}(t)$ averaged over the total field of view (open diamonds) with the local value $M_{x,s}(t)$ [$M_{x,e}(t)$] measured in the central circular area of the square (ellipse) (indicated in Fig. 32) at 0 ps. For better comparison, $M_{x,m}(t)$ was normalized to the same maximum amplitude as the local values. At first glance, the time dependences $M_x(t)$ are close to each other and resemble that of a critically damped oscillation. The simultaneous fluctuations of $M_x(t)$ near 600 and 1500 ps are caused by adjustments of the electron optics and beam injection at the synchrotron.

The local variations of $M_x(t)$ are emphasized in the difference image shown as an inset in Fig. 34b and by the differences shown in Fig. 34c, revealing the true discrepancies between averaged and local magnetization dynamics.

Residual small edge domains that do not participate in the magnetization rotation cause the positive (negative) constant offset of $M_x(t)$ for the ellipse (square). For the ellipse, the difference between local and average magnetization (open circles in Fig. 34c). reveals a broad maximum coinciding with the strong counter clockwise rotation of $M(r,t)$. This behaviour indicates a slower rebound of the magnetization in the centre of the particle that can be explained by the attenuation of the bias field by the in-plane demagnetization field of the ellipse, which reveals a hard axis parallel to the bias field. Contrarily, the difference for the square particle (open diamonds in Fig. 34c) shows an oscillation with a frequency of about 1.7 GHz.

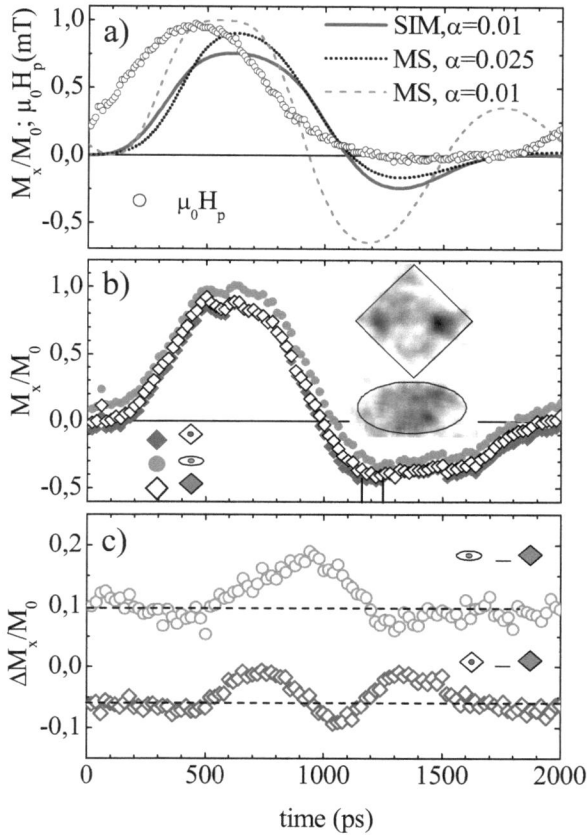

Fig. 34: Dynamic response of a diamond-shaped and an elliptical spin-valve element. (a) Field pulse $H_p(t)$ (open circles) with a repetition rate of 0.5 GHz. Curves show results for the magnetization component $M_x(t)$ predicted by the macrospin (MS) model with low (dashed line) and high (dotted line) damping coefficient. $M_x(t)$ calculated by a micromagnetic simulation[71] (SIM) for the square pattern is shown by the full line. (b) $M_x(t)$ averaged over the complete field of view [$M_{x,m}(t)$, open diamond] and in the central area of the square spin-valve platelet [$M_{x,s}(t)$, full diamond] and of the elliptical particle [full circle] versus time delay. The inset shows a difference image between images acquired at times 160 and 1260 ps. (c) Difference between local magnetization and average magnetization for the central areas of the square (open diamond) and the ellipse (open circle)[56].

The difference image shown in the inset of Fig. 34b relates this frequency to a spin wave mode identified by the two circularly shaped black areas coinciding with the maxima (minima) of the profiles indicated by the dashed vertical lines in Fig. 33a. Because of the presence of this higher-order spin wave mode, the magnetization vector rotates faster in the region of these maxima (minima) compared to the nodes positioned in between, thus leading to a change of rotation direction of the magnetization vector across the square particle for certain delay times. The period of this mode is 600 ps, which corresponds to $f = 1.7$ GHz. For the elliptical particle, such a higher-order mode is not observed. The direction of the magnetization rotation does not change sign across the diameter. The fundamental eigenmode frequency of the square estimated from a similar measurement using smaller and shorter field pulses takes a value of $f = 0.8$ GHz in the field-free time range. Neglecting lateral demagnetizing fields and assuming a macrospin model, the ferromagnetic resonance frequency for exchange biased films is given by

$$2\pi f = \gamma \sqrt{M_s H_A} \qquad (13)$$

with the gyromagnetic ratio γ and the total anisotropy field $H_A = H_{exch} + H_s$ including the exchange bias field H_{exch} and an induced in-plane uniaxial anisotropy field H_s. For the saturation magnetization $\mu_0 M_s = 1.3$ T, we assumed a weighted average of the bulk values for NiFe and CoFe (free layer consists of 2.1 nm NiFe and 1 nm CoFe). Under these conditions, the observed eigenmode frequency of 0.8 GHz corresponds to $H_A = 0.6$ mT, in eq. (13), in agreement with the quasi-static value of H_{exch} derived from the easy axis magnetization curve, see Fig. 21b. Single-domain (macrospin) simulations considering the complete magnetic layer stack suggest that the rotation of the pinned layer can be safely neglected as will be shown later (Fig. 39). The single-domain model has to be adjusted with a high damping constant of $\alpha = 0.025$ and $H_A = 0.6$ mT in order to approximate $M_x(t)$ (see dotted curve in Fig. 34a). The damping coefficient agrees with results reported by Schumacher et al.[39] for the free layer of a very similar spin valve element. Contrarily, a damping constant of $\alpha = 0.01$, which is closer to values reported for the unbiased free layers[40], results in a second maximum of $M_x(t)$ at 1800 ps (dashed curve) that is clearly not observed in our experiment. A full micromagnetic simulation of the square using $\alpha = 0.01$ and replacing the exchange bias stack by a constant field yields a closer agreement with the experimental results (solid curve). It also reproduces the relatively large rebound maximum of $M_x(t)$ in contrast to the macrospin result. The snapshots of the micromagnetic simulation shown in Fig. 23b reveal the characteristic differences between the square and the ellipse in agreement with the experiment: While the magnetization has already rotated back into the equilibrium position in the centre of the square, M still shows a clockwise rotated position toward the left and right corners. In contrast to this behaviour, the ellipse shows a homogeneous magnetization direction along the long axis. Clearly, the shape is responsible for these differences.

Besides the square and the ellipse rectangles measuring 15×10 μm² and 5×10 μm² have been investigated on a 20 μm wide micro stripline. Shape anisotropy points parallel to the micro strip line for the large structure and perpendicular for the small particle. The directions of exchange bias field H_{exch}, pulsed Oerstedt field H_p and the incident light are depicted in Fig. 35c. Both rectangular platelets exhibit an almost uniform magnetization state, which cannot be observed in unbiased Permalloy platelets[24] since for particles with several microns size the Landau-Lifschitz domain configuration is energetically preferred. In contrast to Py, for the pinned multilayers a quasi-single domain state is observed, even for large platelets. Thus, significant stray fields occur due to the pinning field H_{exch}. At higher magnification, a densely packed system of interacting low-angle Néel walls becomes visible which stabilizes itself and forms a buckling state (see Fig. 36).

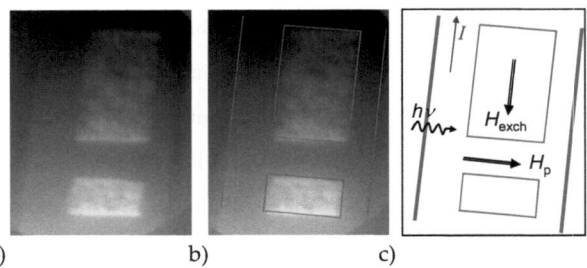

Fig. 35a: PEEM image of two rectangular spin-valve elements on a 20 µm strip line. (b) Same with boundaries of the centre strip line as well as contours of the two rectangles measuring 15 ×10 µm² (upper platelet) and 5 ×10 µm² (lower platelet). (c) Directions of the exchange bias field H_{exch}, the pulsed Oerstedt field H_p and the incident light hv (projected into the drawing plane).

Presumably, the formation of buckling domains is also supported by the grain structure of the underlying Cu stripline. Apparently, the only way for the pinned free layer to reduce energy is the emergence of edge domains (bright horizontal rims in Fig. 36) according to the pole avoidance principle[10]. Due to the demagnetization field a reduction of magnetic charge takes place at the rim of the platelet which is

Fig. 36: Element selective XMCD imaging (Ni L_3 edge) of the free layer of the rectangular spin valve structures reveals buckling domains with small-angle Néel walls. The overall net magnetization appears in the "S" state due to the pole avoidance principle. The red arrows indicate the averaged local magnetization direction at the edges and in the centre regions, forming the "S" state in both cases.

compensated for by the appearance of volume charges. In other words, charges prefer to smear out, to decrease magnetostatic energy. This way both platelets in Fig. 36 appear in the "S" state independent of their orientation to the stripline. As the pulse field H_p is perpendicular to the exchange bias field, it causes an in plane rotation of the magnetization. The initial ground state orientation is essentially parallel to the strip line due to the exchange anisotropy field H_{exch} (Fig. 35c). This becomes clear by analyzing the grey scale values within the regions of the spin valve platelets. Fig. 37 shows an XMCD-PEEM sequence for these elements. As the magnetic field pulse propagates through the strip line, the magnetization rotates coherently out of its initial ground state orientation, falls back into it and "overshoots" to the opposite side after the pulse has passed (see arrows).

The grey scale, giving the component $M_x(t)$ according to eq. (7) is quantified in Fig. 38. It can be deduced that the normalized grey scale intensity averaged over each platelet has a negative component for both particle sizes. After following the magnetic excitation (white appearance of the platelets in Fig. 37), the particles reveal an averaged black XMCD contrast, i.e. $M_x(t)$ turns negative even though the platelet has never been excited in this direction. This behaviour cannot be explained assuming static magnetic behaviour but supports conservation of angular momentum according to the Landau-Lifschitz-Gilbert formalism.

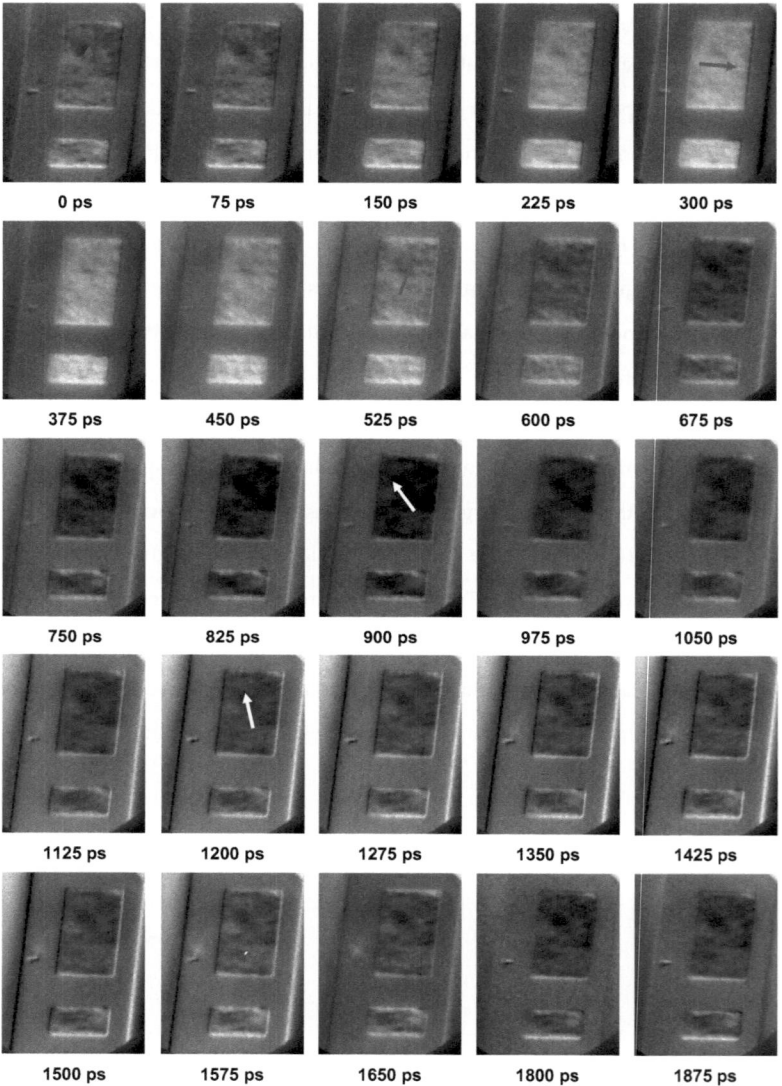

Fig. 37: Series of XMCD-PEEM snapshots showing the dynamics of magnetization in the free layer of two GMR spin valve elements under remagnetization. The upper rectangle measures 15 ×10 µm², the lower 5 × 10 µm². Arrows denote average magnetization directions.

Since the response for fast field pulses are almost critically damped (see 100 ps and 200 ps curves in Fig. 38b and c), fast magnetic switching is possible[39,72]. Similar multilayer stacks have already been implemented successfully as read elements for fast magnetic bits' stray field readout in hard disc drives.

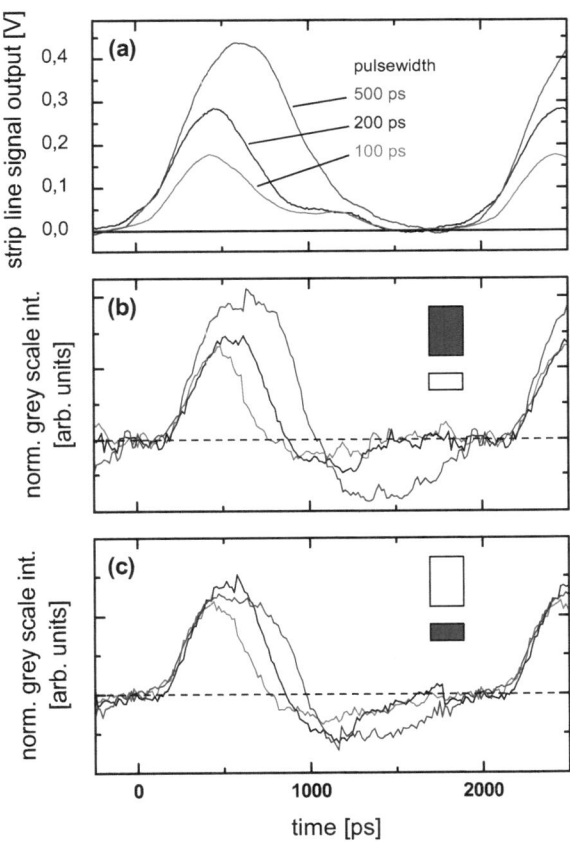

Fig. 38: Dynamic response of two rectangular spin-valve elements extracted from Fig. 37. (a) Stripline output signal for three different pulse widths (100, 200 and 500 ps). Grey scale intensity (normalized) of the 15 ×10 µm^2 structure (b), and 5 ×10 µm^2 structure (c).

The Object Oriented Micromagnetic Framework[73] (OOMMF) has been used to simulate the magnetization dynamics of al three ferromagnetic layers in the GMR spin valve. Standard micromagnetic simulations require a size of the basic cell for the calculation which is small against the size of the platelet but still within the magnitude of the magnetic exchange length. Thus, computational power usually limits the dimension of the simulated platelets. Two models have been compared for platelets with the same aspect ratio as the ones experimentally studied. The macrospin model resembles the quasi-single-domain state of the platelet and assumes coherent rotation. Here, the cell size for simulation has intentionally be set equal to the size of the platelet. In the micro spin model, in contrast, the cell size for simulation has been chosen small enough to appropriately respect the magnetic exchange length ($10 \times 10 \times 3$ nm). In both cases the damping constant for the simulation has been set to $\alpha = 0.1$.

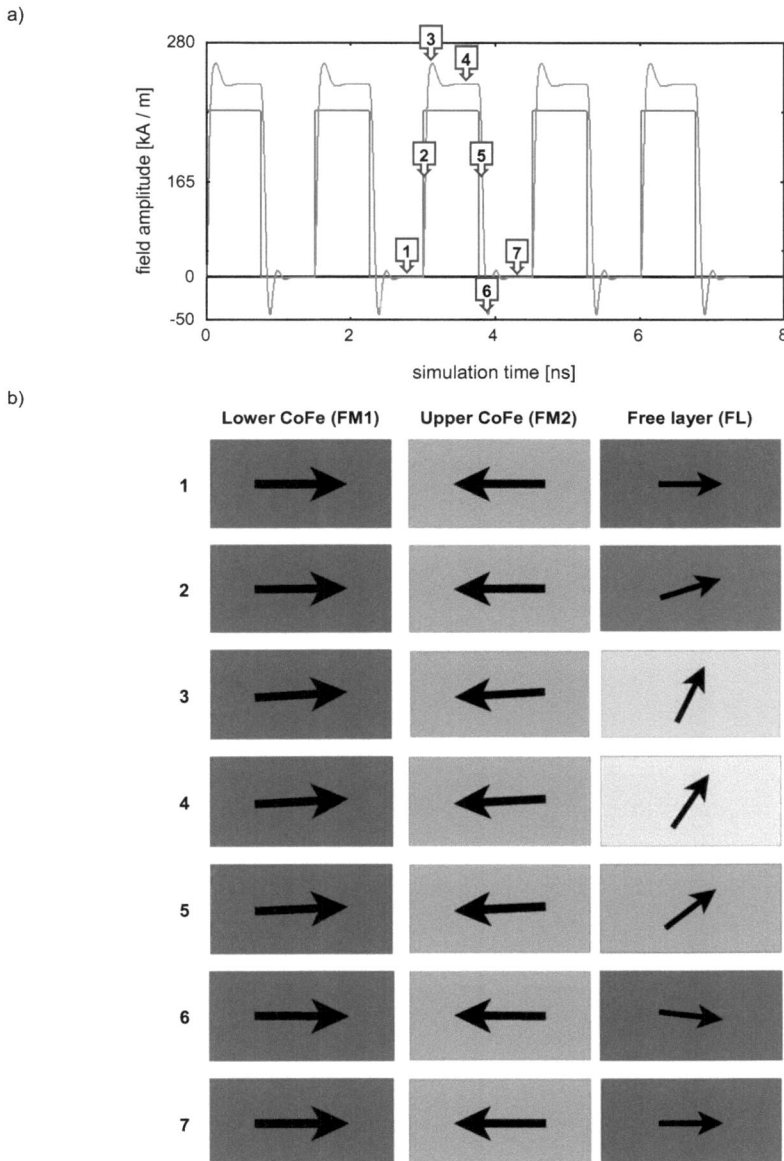

Fig. 39: Micro-magnetic simulation of the magnetization dynamics of a GMR spin valve for a square field pulse (a) assuming a simple macro-spin model. The red and green curves in (a) show the pulse profile and the M_y component, respectively. The arrows in (b) denote the magnetic rotation of each ferromagnetic layer in the spin valve stack according to the time steps 1 to 7. The color code denotes the magnetization direction as defined in (b).

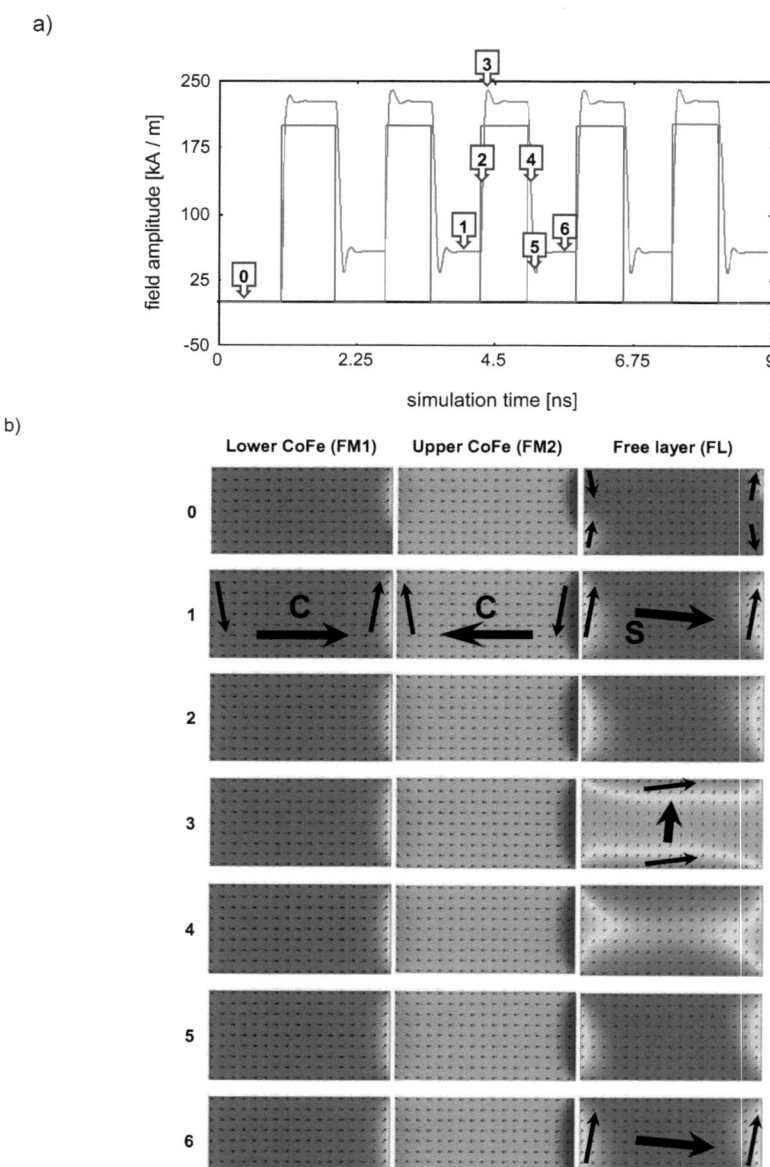

Fig. 40: As in Fig. 39, the red and green curves in (a) show the square pulse profile and the M_y component, respectively. The large and small arrows in (b) denote the magnetic rotation of each ferromagnetic layer in the spin valve stack according to the time steps 0 to 6. The color code denotes the local magnetization direction as defined in (b).

In Fig. 39 and Fig. 40 the results of micro-magnetic simulations with OOMMF using the NIST code of a GMR spin valve are presented. For simplicity, a step like external field with amplitude 20 mT has been applied (red curve) in (a), respectively. We assumed the following material parameters (see Fig. 41):

All three ferromagnetic layers and the non-magnetic spacers are 3 nm thick. The saturation magnetization M_{sat} has been set to 1400 kA/m for free layer and to 2000 kA/m for the hard ferromagnetic layers. The coupling of the lower ferromeagnet to an imaginary antiferromagnet was introduced by a biasing field of 40 kA/m acting on FM1.

The macro-spin model (see Fig. 39) assumes a single domain state, i.e. the magnetization rotation within the platelet is per definition coherent. The magnetic rotation of each ferromagnetic layer in the "mono"-spin model is shown for several selected time steps. As expected the soft magnetic free layer rotates freely into the field direction, whereas the magnetization of the hard magnetic layers rotates only faintly. The rotation of the hard magnetic layers can thus be neglected at our given field pulse amplitudes.

In the micro spin model (see Fig. 40) the free layer reveals edge domains in the "S" state as observed experimentally (cf. Fig. 36). Surprisingly, the simulation however shows, that the underlying hard magnetic layers are both in a "C" state, but with opposite orientations.

The magnetic exchange coupling between the three ferromagnetic layers in the spin valve element can thus be described classically as three rotational objects connected by two springs with different spring constants, where $k_1 \ll k_2$ as depicted in Fig. 41.

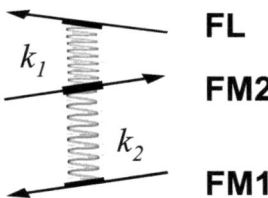

Fig. 41: Spring model illustrating qualitatively the coupled three-layer system, like rotating objects with high moments of inertia, the hard magnetic layers FM1 and FM2 are connected by a spring constant k_2. In the same way the free layer FL is connected to the upper hard magnetic layer with a spring constant k_1. Simulation shows that the rotation of FM1 with respect to FM2 can be neglected because $k_1 \ll k_2$.

The important results of this simulation of the modelled three-layer system are:

- The mono spin model shows almost no response of the hard magnetic layers FM1 and FM2

- The micro spin model shows pronounced edge domains in the "C"-state for the hard magnetic layers (FM1 and FM2). The edge domains in the free layer FL reveal an "S"-state

- The dynamics of the free layer substantially deviates from a coherent response.

5.2. Epitaxial Co Platelets on a Single-Crystalline Mo Strip Line

Fig. 42 shows the time dependence of the field pulse as measured behind the epitaxial strip line on Al_2O_3. Due to the small damping in the $Al_2O_3(11\bar{2}0)$ substrate compared to the Si substrate used for the Permalloy platelets, the field pulse shows an equally sharp leading and trailing edge, reproducing exactly the output of the pulse generator.

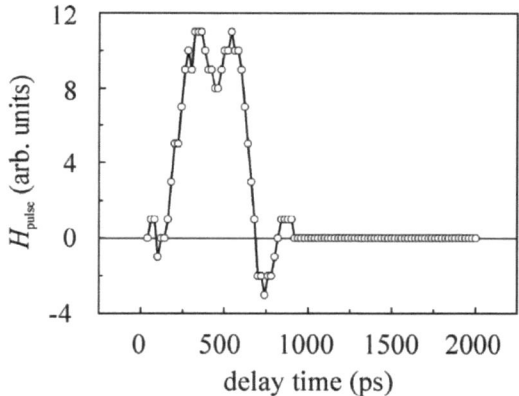

Fig. 42: Temporal profile of the magnetic field pulse $H_p(t)$ on the single-crystalline Mo(110) strip line on Al_2O_3 substrate[50].

The maximum field amplitude calculated from the current signal measured after the pulse has passed the strip line amounts to 6 mT. Co films grown on $Mo(1\bar{1}0)$ show a uniaxial anisotropy with the easy axis along the Mo[110] direction, that in turn is aligned with the strip line axis[53]. The anisotropy constant determined from Kerr magnetometry measured before the final FIB structuring is $K_P = 0.47 \times 10^5$ J/m³ corresponding to an anisotropy field of $\mu_0 H_{aniso} = 50$ mT. The deviation from the results measured for Mo capped $Co/Mo(110)$ films[53] might be attributed to the surface anisotropy induced by the Au capping. Moreover, a decrease of the magnetic anisotropy due to structural damage in the course of the FIB structuring has to be taken into account.

Fig. 43 shows an XMCD-PEEM snapshot sequence with increment 20 ps in the leading edge (starting at 200 ps) of the field pulse in Fig. 42). Unfortunately, the sample was destroyed by electrostatic discharge after image (g) was taken.

The uniaxial anisotropy defines the magnetization direction at equilibrium along the strip line. A homogeneous magnetization state as shown in the Fig. 43a is typical for these epitaxial grown platelets. The dark colour of the platelet in Fig. 43a indicates a magnetization component antiparallel to the incident X-ray beam before the field pulse arrives (see arrow). During the steep leading edge of the field pulse we observe a reversal of the XMCD asymmetry revealing a change of sign of the horizontal magnetization component. Consequently, the magnetization has rotated by at least 45° in Fig. 43g as indicated by the arrows on the right. A quasi-static field of 6 mT would only cause a small rotation of 7° for the given field and anisotropy.

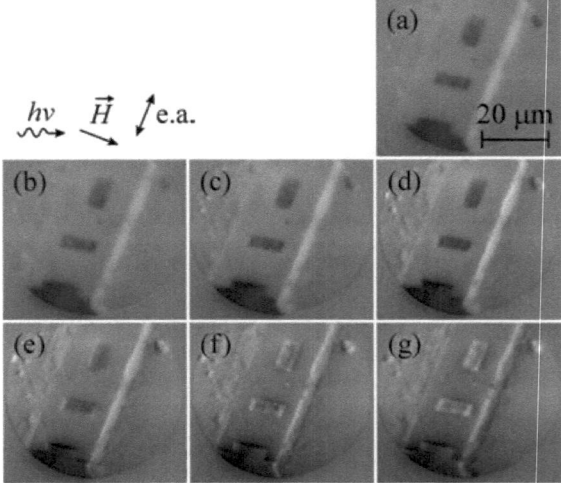

Fig. 43: XMCD-PEEM snapshot sequence of epitaxial Co platelets on a single-crystalline Mo(110) strip line on Al_2O_3. Sequence (a-g) shows the change of magnetization state with time steps of only Δt = 20 ps. The series was taken in the rising edge of the field pulse (between 200ps and 320ps in Fig. 42). Sizes of Co particles on the strip line are 5 × 10 μm^2 and 10 × 5 μm^2 [Ref.[50]]

This unexpected large response might be attributed to a decreased magnetic anisotropy deviating from the measurement on the unstructured area of the sample. The Co elements have been structured by focused ion beam (FIB) milling. We know from other experiments[65] that the anisotropy can change upon ion bombardment. In addition, we observe a non-uniform response of the magnetization. Close inspection of Fig. 43f-g reveals bright rims, i.e. a larger positive asymmetry (positive horizontal magnetization component) at the left and right edges of rectangular elements. This observation indicates a stronger dynamic response of the magnetization at the edges where the field enters and leaves the ferromagnetic particles. This is quite surprising since one would expect a stronger response just at the other edges because of the demagnetizing field occurring at the poles of the magnetic sample. On the other hand the demagnetizing field is negligibly small for the ultrathin film investigated here and a lateral inhomogeneous magnetic anisotropy due to the structuring might be the reason for this observation.

From the MOKE loop in Fig. 27 it can be deduced, that the coercivity field necessary to magnetically saturate these platelets is at least 25 mT. The field pulse generated by the Oerstedt field was probably not sufficient to drive the magnetization into complete saturation.

The sequence in Fig. 43 proves a time resolution better than 20ps. Between (d) and (e) and even more pronounced between (e) and (f) and (f) and (g), we observe marked changes of the asymmetry patterns. This result is in good agreement with our earlier measurements on permalloy particles[24], where we found a time resolution of 14 ps. Given the photon pulse width of approximately 5 ps in the low-α mode, the main contribution stems from jitter in the electronic set-up, i.e. bunch-marker from BESSY and magnetic pulse generator.

5.3. Propagating Magnetic Eigenmode in Ultra-Thin Py Ellipse

Fig. 44 shows a sequence of snapshots of XMCD-PEEM images taken with 25 ps time increment for an elliptical Permalloy particle with 3 nm thickness. As described in chapter 4.3, the Py has been deposited by Ion Beam Deposition. Field pulses of about 500 ps width and an amplitude of 1.5 mT were applied, again with a repetition rate of 500 MHz. Since the maximum of the applied field pulse is larger than the quasi-static value that would be needed to saturate the particle, we assume that the observed magnetization dynamics does not depend on the initial state before the pulses were applied. Note that in this experiment the pulse field is oriented perpendicular to the photon beam, see arrows in Fig. 44.

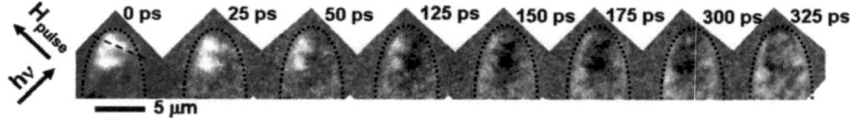

Fig. 44: XMCD-PEEM image series of a propagating spinwave in the upper part of an elliptical thin-film structure (semi-axes 6 μm × 12 μm, 3 nm thick). The dashed line in the 0 ps image denotes the position of the line scan shown in Fig. 45[70].

The snapshots were taken with increasing delay time between magnetic field pulse and the bunch marker signal of the storage ring. Here, the delay time $t = 0$ ps roughly corresponds to the maximum of the field pulse. We infer absolute values from flux-closure magnetization states of smaller particles before the field pulse measured in the same run. This calibration results in the fact that at $t = 0$ ps the magnetization close to the upper boundary of the elliptical particle is oriented nearly perpendicular to the orientation of the pulse field in contrast to the expectation for a quasi-static behaviour. This observation is in course agreement with a dynamic simulation.

In order to view the spatiotemporal behaviour of the XMCD contrast quantitatively, a line scan along the dashed line in the image series in Fig. 44 has been plotted as function of time. The result is displayed in Fig. 45. All XMCD asymmetry profiles shown in Fig. 45

deviate considerably from conventional domain wall profiles observed during quasi-static remagnetization processes.

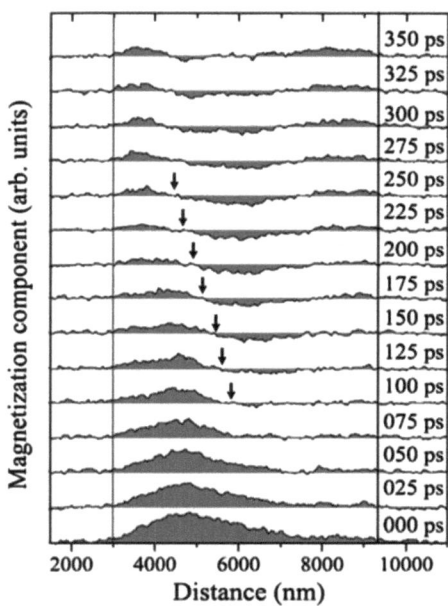

Fig. 45: Profiles of the set of XMCD-images along the line shown in the left image if Fig. 42 for increasing time delay. The profiles indicate the magnetization component $M_x(t)$ along the incident light. The delay times cover the time range during the maximum of the field pulse. Arrows denote a propagating phase front. The vertical lines denote the boundaries of the particle[70].

The domain wall width for quasi-static conditions is determined exclusively by exchange constant and magnetic anisotropy resulting in values of the order of 200 nm for Permalloy, cf. Néel wall widths in Fig. 31. The profiles shown in Fig. 45 instead resemble a wave pattern with a wavelength comparable to the dimensions of the particle, i.e. few micrometers. The shape of the wave pattern evolves during increasing time delay from an asymmetric maximum with only positive values at $t = 0$ ps (bottom profile) to a symmetric wave with two maxima and a minimum in between at $t > 250$ ps.

Considering a fixed position on the particle we observe an oscillating behaviour in time for the magnetization component. At an adjacent position a similar oscillation in time occurs, however, with a phase shift. This is a clear indication of a travelling spinwave (comp. Fig. 8). On the other hand, the boundary condition (boundaries denoted as vertical lines in Fig. 45) cannot be neglected for the description because of the long wavelength comparable to the particle dimensions. This case can be seen as an intermediate case between free travelling spin waves and standing waves as described above. Moreover, the rotation angle of the magnetization vector is large compared to thermally excited spin waves, which will certainly influence the phase velocity of spin waves because they are inherently non-linear.

For delay times between $t = 100$ ps and $t = 250$ ps the shape of the spin wave is almost stable and the wave travels with a constant velocity. The velocity can be inferred from the motion of a point with constant phase of the wave, i.e. phase velocity. We use for this purpose the zero crossing as indicated in Fig. 45 by the arrows, resulting in an appearing phase velocity of $v_{ph} = 8100$ m/s. This value is clearly higher than values typically reported for the velocity of domain walls (up to 100 m/s). Even the recently reported much larger values of 1500 – 2000 m/s[74,75,76] remain smaller compared to the velocity determined here. Since the motion with constant velocity is relatively short and also does not take place in a homogeneous film the comparison to free spin waves has to be done with some care. The change of the shape also indicates that the phase and group velocity are different. However, similar values as in our case were found in homogeneous permalloy films for the group velocity of a magnetostatic surface wave[77]. The unexpected high domain wall speed could be ascribed to the very low thickness of the Py layer (3 nm), along with the very good quality and high purity due to IBDS preparation technique.

5.4. Magnetization Dynamics of Ultra-Thin Py Rings

In this experiment again the incident light lies in the horizontal plane. The coplanar waveguide lies in this plane, too, as depicted in Fig. 46. Areas with vertical local magnetization vector appear grey whereas regions with horizontal component appear black and white, respectively. A ring in vortex state therefore features grey sides left and right, a black top and white bottom pole (see Fig. 4a).

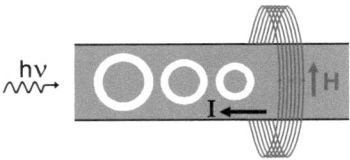

Fig. 46: Horizontal orientation of centre stripe of the wave guide, equipped with Py rings (white) of 3 nm thickness and different sizes. The plane of photon incidence lies horizontal as well. The Oerstedt field H_p (perpendicular to the current I) is oriented vertical in the region of the ring structures.

The magnetic emergence of a ring in the onion state (see Fig. 4b) in X-PEEM resembles a lifesaver. As a metastable state, the onion configuration is accessible from magnetic saturation (see Fig. 4c) in a reproducible way. With decreasing ring diameter however, the energetic difference between the onion state and the flux-closure vortex state decreases. It was observed during the experiment that rings smaller than 6 μm outer diameter undergo spontaneous transitions from onion to vortex state and vice versa.

An IBD-Permalloy ring of 3 nm thickness with inner diameter of 6 μm and outer diameter 15 μm in the onion state is shown in Fig. 47. The left columns show a micromagnetic simulation, the XMCD-PEEM image sequence is shown in the right columns. Initially ($t = -100$ ps) the ring is in the onion state. By applying a magnetic field in vertical direction the ring gets magnetically saturated ($t = 900$ ps in the simulation; 690 ps in experiment) and comes back into the onion state ($t = 1700$ ps).

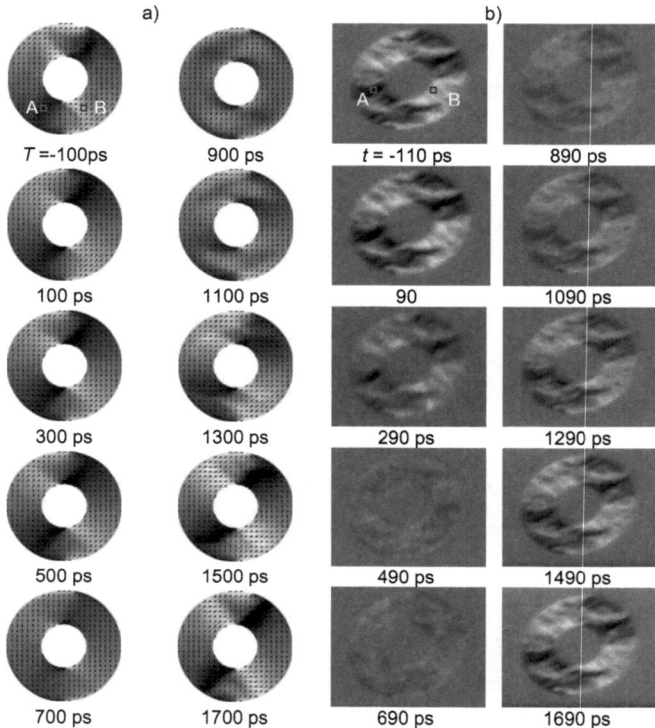

Fig. 47: Magnetization dynamics of a permalloy ring of 3 nm thickness with inner diameter of 6 μm and outer diameter 15 μm. Column (a) shows micromagnetic simulated reaction on the field pulse. Column (b) shows the XMCD-PEEM image sequence. The component of the magnetization vector $M_x(t)$ is obtained from the greyscale level in the regions A and B, respectively.

The graph in Fig. 48a shows the measured of the field pulse profile during this experiment. An Oerstedt field pulse with 1.5 mT amplitude was achieved at the 20 nm wide micro stripline. The normalized x-component is obtained from the grey scale level of the areas of interest A and B denoted in Fig. 47 (first images in (a) and (b)). Fig. 48 shows the value M_x/M_s for simulation (b) and experiment (c). As the magnetization direction in the selected micro regions turns from horizontal to vertical with opposite sign, however, it overshoots the twelve (six) o'clock position in theory and experiment.

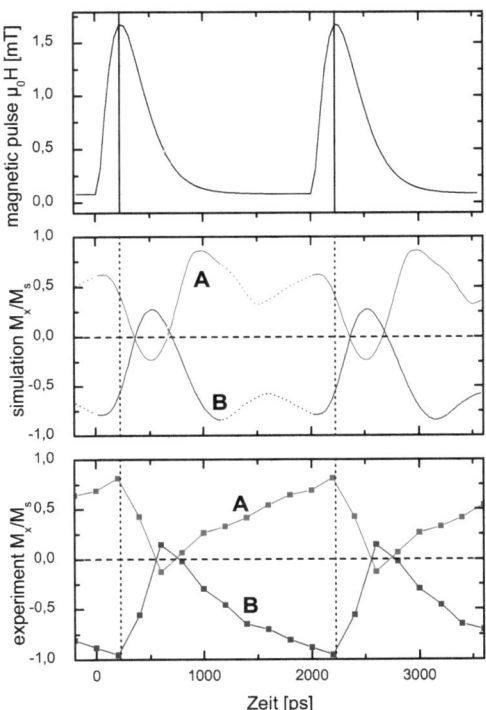

Fig. 48: Local response in micro areas A and B (denoted in Fig. 47) for a magnetic field pulse of 1.5 mT (a). Results of micromagnetic simulation (b) and experiment (c) for the horizontal component of magnetization $M_x(t)$ obtained from grey scale intensity in the image sequence.

The simulation assumed a ring with the outer diameter of 5 μm, a saturation flux density of $B_{sat} = 0.8$ T and a damping coefficient of $\alpha = 0.005$. The cell size for the simulation was 5 nm².

Again, this effect cannot be explained by quasi-static rotation phenomena but originates in the conservation of angular momentum. The agreement with the experiment is fairly good concerning the local response in Fig. 48. The experimental XMCD contrast images appear more "grainy" because of local structural inhomogeneities and a possible system of small-angle Néel walls as also observed for the spin valve sample (see Fig. 36).

5.5. Dynamic Response of 90° Néel Domain Walls

The effect of the magnetic field pulses on the thicker Permalloy micro patterns described in chapter 4.4 is shown in Fig. 49a-h, which compiles a sequence of images taken with a time increment of $\Delta t = 10$ ps. The direction of photon incidence (projection onto the sample surface) and the pulse field $H_p(t)$ are directed from left to right (parallel to the short edge of the particles) and bottom-up (parallel to the long edge of the particles), respectively, as given in the inset.

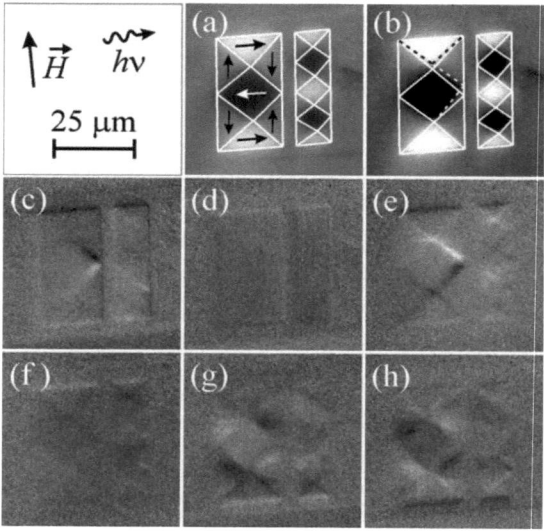

Fig. 49: XMCD-PEEM image sequence of rectangular Permalloy particles taken during the steep rising edge of the field pulse at $t = 100$ (a) and 110 ps (b) as well as difference images calculated from image pairs taken at $t = 110$ (c), 120 (d), 130 (e), 140 (f), 150 (g) and 160 ps (h) and $t^* = t - \Delta t$, $\Delta t = 10$ ps. Directions of light incidence hv (projection on the sample surface) and pulse field H_p are indicated in the inset[50].

Due to the alignment between the local magnetization and the direction of light incidence, only three distinct contrast levels (black, white and gray) appear. The XMCD-PEEM image in Fig. 49a shows the domain structures of the particles under study at $t = 100$ ps, i.e. at the onset of the field pulse. These structures are schematically shown by solid lines with arrows denoting the local magnetization direction. They are very similar to the initial flux closure (Landau) domain patterns of these structures. Following the evolution of this pattern in time, we observe that the domain structure is visibly affected, i.e., the magnetization distribution is locally deformed. The domain boundaries shift significantly within the next 10 ps ($t = 110$ ps), see Fig. 49b. The dotted lines in Fig. 49b mark the new positions of the domain boundaries, as given by the contrast changes from black to white. These field-induced changes may be emphasized, if the domain pattern from Fig. 49a is subtracted from the image in Fig. 49b (see Fig. 49c). Fig. 49c-h show a series of difference images calculated from image pairs taken at subsequent points in time $t = 110$ (c), 120 (d), 130 (e), 140 (f), 150 (g), 160 ps (h) and $t^* = t - \Delta t$ ($\Delta t = 10$ ps). The fact that these difference XPEEM images exhibit distinct features indicates two important findings. First, the actual time resolution is significantly smaller than the chosen step width. Second, the observed changes are reversible and appear during all (or at least most) of the cycles in our stroboscopic experiment. The results of measurements given in Fig. 49a-h correspond to the current pulse leading-edge (100 ps $\leq t \leq 160$ ps), denoted as area I in Fig. 14a.

Fig. 50: XMCD-PEEM image sequence of Permalloy particles taken in the region of the pulse maximum at $t = 200$ ps (a) and difference images corresponding to the pairs $t = 220$ (b), 240 (c), 260 (d), 280 (e), 300 (f) and 320 ps (g) and $t^* = t - \Delta t$, $\Delta t = 20$ ps. Line AB denotes the 90° Néel domain wall. Near the line AB the contrast of the region marked by the arrow varies (b-g)[50].

A closer inspection of the data in Fig. 49 reveals that the changes during the time sequence are mostly confined to the domain boundaries. The larger particle responds stronger than the smaller one. In addition, the changes do not follow a simple trend, but rather an almost oscillatory pattern, although we are still on the rising edge of the pulse. This shows up very clearly when comparing Fig. 49c and e, in which the contrast indicating the response of the domain boundaries is reversed.

Therefore, the behaviour of the magnetization at the domain boundaries is more complex and cannot be described by a simple linear motion, i.e. a domain wall shift as in the case discussed in Ref.[24]. It further proves the significance of the higher Fourier frequencies contained in the steep rising edge (see Fig. 14b). In the further course of the field pulse changes occur at different locations in the magnetization pattern, as depicted in the sequence in Fig. 50a-g. In the following, we will concentrate on the dynamic behaviour in the region close to the maximum of the field pulse (Fig. 14, area II). Near the flat maximum the overall response is weaker than during the rising edge which again points on the importance of the Fourier components with a strong damping of these high-frequency modes in the particle. In this sequence the smaller particle shows a contrast whereas the larger one yields practically no response. Fig. 50b-g demonstrates the changes in the magnetization distribution in the area containing the 90° Néel wall marked by the line AB in (a) between the domains in the Landau pattern. The difference XMCD-PEEM images taken at time $t = 220$ (b), 240 (c), 260 (d), 280 (e), 300 (f), 320 ps (g) and $t^* = t - \Delta t$ ($\Delta t = 20$ ps), map the evolution in time. As can be seen, close to the line AB the contrast in the region marked by the arrow varies distinctly. This is caused by a change of the magnetization direction in the region adjoining the 90° Néel domain wall AB. The dynamics of this contrast change is compiled in Fig. 51a.

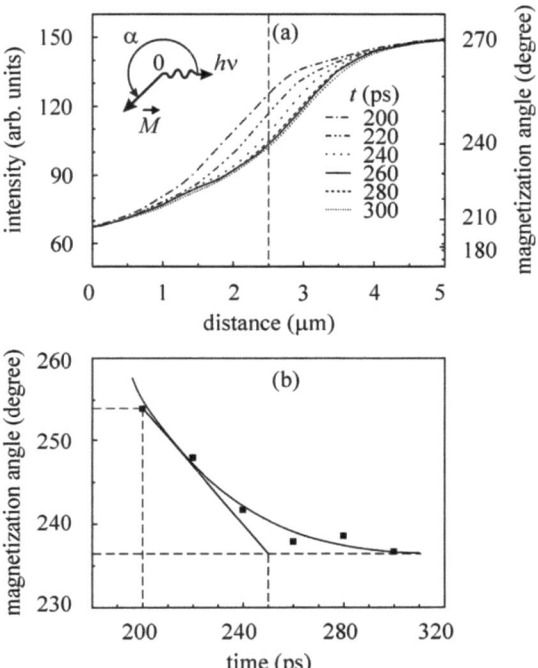

Fig. 51: (a) Intensity profiles j of the grey values in Fig. 48 along the dotted line marked in Fig. 50a for the XMCD asymmetry images taken between time $t = 200$ and 300 ps. The right ordinate gives the corresponding angle dependence $\alpha(x)$. The magnetization angle α is counted from the illumination direction (its projection on the particle plane) as depicted in the inset. The dashed vertical line denotes the centre of the region marked by the arrow in Fig. 50b-g and corresponds to the position of line AB in Fig. 50a. (b) The angle of the local magnetization direction $\alpha(t)$ in the region marked by the arrow in Fig. 50b-g as function of time[50].

It gives a series of intensity profiles $j(x)$ through the region under study extracted from the XMCD-PEEM images taken between $t = 200$ and 300 ps. These profiles are recorded along the dotted line being perpendicular to line AB (Fig. 50a). According to eq. (7) the grey level in the XMCD contrast is proportional to $\cos\alpha$, where α is the angle between the directions of magnetization in the particle plane and projection of direction of photon illumination onto this plane. The angular dependence $\alpha(t)$ (Fig. 51a, right hand scale) can be directly extracted from the intensity profiles $j(x)$ (left hand scale). The extreme values $\alpha = 180°$ and $270°$ on the right-hand scale correspond to the grey levels for the magnetization direction being antiparallel and perpendicular to the projected photon impact direction, respectively.

We clearly see that the profiles $\alpha(x)$ in Fig. 51a shift to the right and deform as t increases. This is a further indication that we cannot interpret the series of curves in Fig. 51a in terms of a quasi-static movement of the 90° Néel domain wall AB.

The vertical dashed line in Fig. 51a corresponds to the centre of the region marked by the arrow in Fig. 50b-g. From the variation of $\alpha(x)$ along this line, we can extract the time dependence of the angle $\alpha(t)$ as shown in Fig. 51b. An inaccuracy in the determination of the angles can be caused by the fact that not all of the 10^{13} cycles of the passage of the magnetic field pulse are characterized by complete repeatability, or there is a non-negligible out-of-plane component of the transient magnetization. It means that the real curve $\alpha(t)$ can be shifted to somewhat higher angles. As is seen in Fig. 51a, the monotonous character of the shift of the intensity profile lines $j(x)$ is disturbed from the left to the right with increasing t. This can be induced by the above-mentioned effect as well as by the wave character of spin wave propagation. The latter reason is the most probable and it can explain an oscillating variation of the position and inversion of the contrast of the region marked by the arrow in Fig. 50b-g.

Any change of the magnetic configuration on such a short time scale is accompanied by the emission of spin waves. The top panel of Fig. 52 shows the result of a simulation of spin waves generated at a 90° Néel domain wall. This computer simulation was realized with the help of the program OOMMF for a Permalloy particle of rectangular shape with aspect ratio 2:1. The magnetic field pulse profile was chosen similar to the measured one shown in Fig. 14a. The local directions of magnetization in Fig. 50 are shown as small arrows.

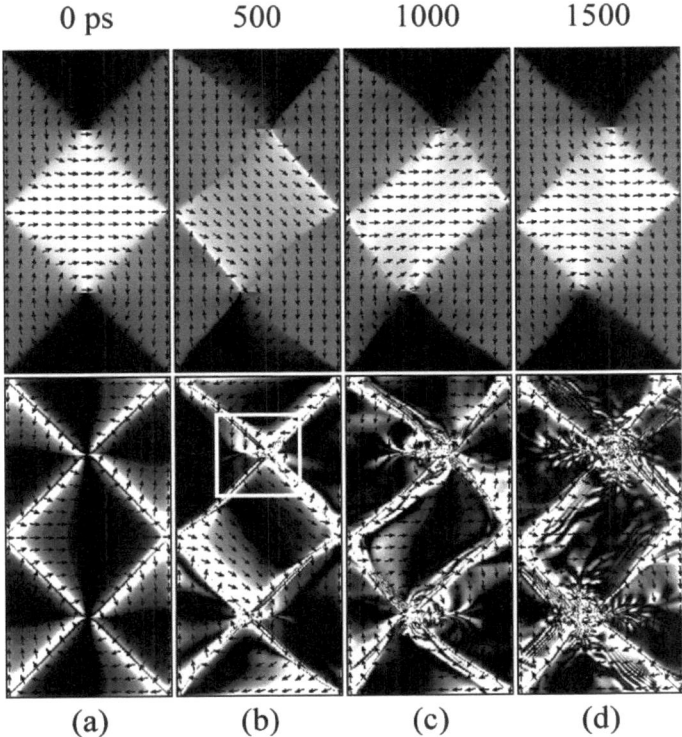

Fig. 52: Emission of spin waves from 90° Néel domain walls (simulation using the NIST OOMMF code[73]). Top row: Micromagnetic simulation showing the time evolution of the magnetization (bright areas are magnetized to the right, dark areas to the left) in a Permalloy particle with linearly reduced dimensions (4 μm × 2 μm × 10 nm) for delay times t = 0 (a), 500 (b), 1000 (c) and 1500 ps (d). Bottom row: Corresponding divergence of the magnetization on the surface of the particle[50].

The simulation was carried out with a step width of 10 ps, a cell size of 40 nm, B_{sat} = 1 T and α = 0.008. The results for time steps of 500 ps are given in the top row of Fig. 52. From the comparison of adjacent images, one sees distinct shifts of 90° Néel domain walls and deformations of the domains themselves. They are close to experimentally observed shifts. The stronger the local magnetic field, the larger the shifts. The first ones are visualized in the bottom row of Fig. 52 as gray levels denoting the divergence of the magnetization. These variations are shown with the step of 50 ps in more detail for the area marked by the white square in Fig. 52 as difference images of magnetization divergence in Fig. 53.

Considering the high Fourier frequency components in our experiments, we can assume that particular spin wave modes are driven almost into a resonant behaviour, comparable, for example, to a parametric oscillator. It has also been shown recently that spin waves may propagate through domain walls[78]. The overlap of these spin wave modes and the non-collinear magnetization distribution in the domain wall itself give rise to a local variation of the magnetization and may explain the deformation of the profiles observed in Fig. 51a. We can try to estimate the period of this mode from Fig. 51b. It can be determined from the condition that the magnetization direction in the spin wave should change by $\alpha = 360°$ during a full period. The period T is equal to about 900 ps (the frequency is about 1 GHz) because, as is seen from Fig. 51b, the angular velocity of the magnetization precession reaches a maximum slope of 0.4° in 1 ps.

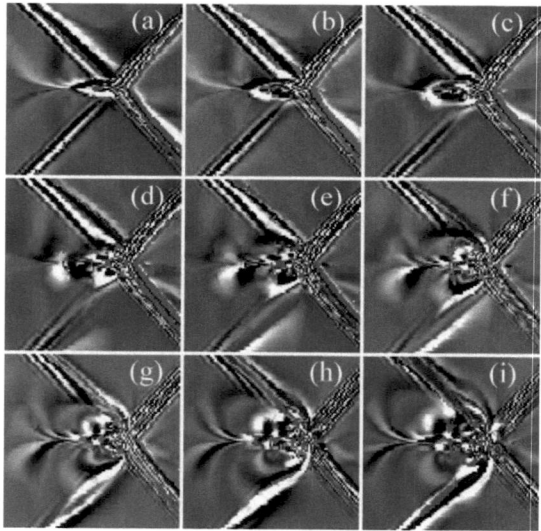

Fig. 53: Difference images of magnetization divergence for the area designated by the white square in Fig. 52 at time t = 550 (a), 600 (b), 650 (c), 700 (d), 750 (e), 800 (f), 850 (g), 900 ps (h) and 950 ps and $t^* = t - \Delta t$, Δt = 50 ps[50].

The characteristic time of the magnetization reversal ($\alpha = 180°$) is $T/2$, i.e. about 450 ps. A movement of a 90° Néel domain wall means that in the region of the domains adjacent to the wall, on one side the magnetization direction turns by the angle of $\alpha = 90°$ and becomes parallel to the domain magnetization direction on its other side. It means that the second domain grows on the expense of the first domain, and in such a manner the domain wall moves. The characteristic time needed for this 90° rotation process is $T/4$, i.e. about 225 ps. This result is close to the data available in the literature[5]. For example, the study of the magnetization dynamics of similar samples performed by means of MOKE (square Permalloy particles with edge lengths of order of a few microns and 18 nm thick) showed that the intrinsic frequency of the magnetization reversal corresponds to modes of a few gigahertz[5]. Besides, the frequency depends an the particle dimensions dependence (it increases as the particle lateral size decreases). The velocity of the 90° Néel wall AB is determined from the shift along the abscissa of the profiles in Fig. 51a taken with time increments of Δt = 20 ps. The maximum velocity is 15.000 m/s.

Chapter 6

Conclusions and Outlook

Conclusions and Outlook

A stroboscopic pump-probe technique has been applied for the investigation of the dynamics of ferromagnetic particles responding to ultrafast magnetic field pulses (resulting from a current pulse through a micro stripline). The time resolution is determined by the photon pulse width as well as the jitter of the bunch-marker and current pulse generator output. In the performed measurements in the low-alpha multi-bunch mode of BESSY the photon pulse width was < 2 ps and the jitter was about 15 ps. The resulting total time resolution was thus essentially restricted by the jitter. It is determined by the pulse electronics, the length of the connecting cables etc. It means that a time resolution in the 10 picosecond range can be achieved with the help of the applied technique. Thus, the observation of fast processes in the gigahertz range became only possible due to the joined effort of producing ultra short X-ray pulses at BESSY II (operated in low alpha mode) and setting up a time-resolved PEEM experiment. In it's standard operation mode, BESSY II delivers photon pulses of typically 50 ps width. Since improving the knowledge in the field of magnetization dynamics becomes increasingly important, stroboscopic PEEM exploiting X-ray magnetic circular dichroism (XMCD) is on its way to contribute significantly to this field.

One main system studied was a multilayer spin valve structure fabricated by industrial standards in the wafer fab of Sensitec GmbH, Mainz. The silicon wafers were processed in thin film technology suited for manufacturing high performance magnetoresistive sensors. 125 mm wafers have been processed in a class 100 clean room. Deposition and structuring of the layer stack of the coplanar waveguide equipped with GMR spin valves was done with standard lithography at λ = 365 nm and dry etching and ion milling techniques.

In conclusion, we find that the magnetization dynamics of the free layer of a GMR spin valve stack deviates significantly from a simple phase-coherent rotation as expected from the macrospin model. The dynamic response of the free layer is a superposition of an averaged critically damped precessional motion and localized spin wave modes, which depend on the shape of the micro pattern. A micromagnetic simulation qualitatively reproduces the observed spin wave mode for the square platelet. In principle, higher-order spin wave modes provide an additional efficient channel for energy dissipation and thus should result in a higher effective damping coefficient (α = 0.01), as observed in our experiments. First, we found the damping coefficient to be independent of the shape of the spin-valve element, thus favouring the model of nonlocal magnetization damping. However, building the difference in the magnetic rotation from the central region of the diamond shaped (5 × 5 μm) platelet and the outer rim reveals a standing spin wave with a period of $T = 600$ ps, which corresponds to a frequency of $f = 1.67$ GHz.

Micromagnetic simulation suggests that the reaction of the hard magnetic layers in the spin valve stack can be neglected and that a simple monospin model can describe neither sufficiently the observed spin wave modes nor the occurrence of edge domains. The strong exchange coupling leads to the complementary formation of "C" states in the lower and the upper ferromagnet. As expected, the stray field of these compensate at the free layer. Due to the weak coupling of the free layer to the upper ferromagnet, the edge domains in the free layer form an "S" state to minimize stray field H_{demag}.

For quantitative comparison with these coupled-layer structures, we investigated single film structures of Permalloy (thickness of 40 nm) on the central strip line of a coplanar waveguide. We found an induced magnetic moment in a rectangular platelet oriented perpendicular to an exciting oscillatory magnetic field, when the system is excited just below the resonance frequency. The magnetization distribution adapts itself to increase the energy dissipation and thus causes an overall increase of entropy. Above a threshold, the near-resonance spin wave mode exerts an effective force perpendicular to the central 180°-Néel wall that is balanced by the restoring force of the stray field energy.

In ultrathin elliptical Py particles (semi axes 6 μm × 12 μm, 3 nm thick) oriented with the long axis at an angle of 45 degrees with respect to the exciting field, we observed travelling spin waves. High precision in time domain and a snapshot series with 25 ps time increment ensured sufficiently high time resolution. The phase front of the spin wave with large precessional angle propagates with a velocity of 8.100 m/s, i.e. much faster than typical domain walls in Permalloy. Velocity and change of wave shapes indicate that the observed phenomena are closely related to spin waves in homogeneous media.

Rings made of ultra-thin Py (3 nm thick) appear in the so-called "onion state" since the experiment allows to access this meta-stable state from magnetic saturation. In this state a net flux traverses the ring but within the ring flux-closure like domains are formed. The field pulse drives the structure into saturation. However, that state is energetically unfavourable and decays rapidly (within 400 ps) into the onion state after the field pulse has passed. The micromagnetic simulation of a 5 µm sized ring assuming $B_{sat} = 0.8$ T and a damping coefficient $\alpha = 0.005$ qualitatively and quantitatively reproduces the observed over-critical damping.

The high time resolution allowed us to study the dynamics of the deformation of 90° Néel walls in Py structures with a Landau flux closure pattern due to the action of a magnetic field pulse on these particles. The characteristic time for magnetization reversal estimated from this is half of the precessional period $T/2 = 450$ ps. The domain wall is displaced out of equilibrium during action of the field pulse with a maximum velocity of 15.000 m/s, i.e. much faster than typical "quasi-static" wall velocities.

We succeeded to grow an epitaxial Co (0001) platelet on an epitaxial Mo(110) stripline, that was deposited and structured on a single-crystalline sapphire ($11\bar{2}0$) substrate. A homogeneous magnetization state was observed for the epitaxial grown platelet. Under excitation with a 6 mT field pulse of 500 ps pulse width, the two rectangular platelets with 0 and 90° orientation to the micro strip line show an unexpected large response, which might be attributed to a decrease of anisotropy due to structuring the sample. The magnetization has rotated by at least 45°, whereas 7° are expected for given field and anisotropy. The asymmetry is higher at the short edges of the structures, indicating a stronger dynamic response at the edges where the field enters and leaves the magnet.

Finally, a variety of ferromagnetic systems were successfully investigated with TR-PEEM using XMCD. The method has proven to be suitable for the investigation of spin wave phenomena in micron-sized structures. This is due its adequate positioning in the landscape time and lateral resolution.

Chapter 7

Attachments

7.1. List of Publications

Publications related to the topic of this Thesis

F.Wegelin, D.Valdaitsev, A.Krasyuk, S.A.Nepijko, G.Schönhense, H.J.Elmers, I.Krug, C.M.Schneider. *Magnetization dynamics in microscopic spin-valve elements: Shortcomings of the macrospin picture.*- Phys.Rev.B, **76** (2007) No.13, 134410/1-4.

F.Wegelin, A.Krasyuk, H.J.Elmers, S.A.Nepijko, C.M.Schneider, G.Schönhense. *Stroboscopic XMCD-PEEM imaging of standing and propagating spin wave modes in permalloy thin-film structures.*- Surf.Sci., 2007, **601**, No.20, 4694-4699.

F.Wegelin, A.Krasyuk, D.A.Valdaitsev, S.A.Nepijko, H.J.Elmers, G.Schönhense, C.M.Schneider. *Magnetization dynamics in polycrystalline permalloy and epitaxial Co platelets observed by time-resolved photoemission electron microscopy* - physica status solidi B, 1-7 (2009) / DOI 10.1002/pssb.200844486

G.Schönhense, H.J. Elmers, A.Krasyuk, F.Wegelin, S.A.Nepijko, A.Oelsner, C.M.Schneider. *Transient spatio-temporal domain patterns in permalloy microstructures induced by fast magnetic field pulses.*- Nuclear Instruments and Methods in Physics Research B, **246** (2006) No.1, 1-12.

A.Krasyuk, F.Wegelin, S.A.Nepijko, H.J.Elmers, G.Schönhense, M.Bolte, C.M.Schneider. *Self-Trapping of Magnetic Oscillation Modes in Landau Flux-Closure Structures.*- Phys.Rev.Lett., **95** (2005) No.20, 207201/1-4.

V.E.Demidov, D.I.Kholin, S.O.Demokritov, B.Hillebrands, and F.Wegelin, J. Marien. *Magnetic patterning of exchange-coupled multilayers.* Appl. Phys. Lett. **84** (2004) No. 15

C.M.Schneider, I.Krug, M.Müller, F.Matthes, S.Cramm, F.Wegelin, A.Oelsner, S.A.Nepijko, A.Krasyuk, C.S.Fadley, G.Schönhense. *Investigating spintronics thin film systems with synchrotron radiation.*- Synchrotron Radiation in Natural Science (Bulletin of the Polish Synchrotron Radiation Society), **7** (2008) , No.1-2, 66-68 Biuletyny PTPS (Polskie Towarzystwo Promieniowania Synchrotronowego), ISSRNS 2008.

A.Krasyuk, F.Wegelin, S.A.Nepijko, H.J.Elmers, G.Schönhense, I.Mönch, H.Vinzelberg, C.M.Schneider. *Dynamic of 180° Néel walls in two-dimensional permalloy particles observed via picosecond time-resolved photoemission electron Microscopy.*- Ukr. J. Phys., **54** (2009) No.1-2, 170-175 (Ukr. Fiz. Zhurn. **54** (2009) No.1-2, 177-182).

Annual Report and Proceedings

F. Wegelin, A. Krasyuk, D. Valdaitsev, S.A. Nepijko, H.J. Elmers, G. Schönhense, I. Krug, C.M. Schneider. *Free-Layer Dynamics of a Synthetic Spin Valve with Antiparallel Pinning.-* BESSY Annual Report 2005, edited by K.Godehusen. Berlin: published by Berliner Elektronenspeicherring-Gesellschaft für Synchrotronstrahlung m.b.H. (BESSY), 141-142.

F. Wegelin. *Zeitlich aufgelöste Abbildung von Ummagnetisierungsprozessen.* – Proceedings of the 9. Symposium on magnetoresistive Sensors and magnetic Micro Systems (2007). published by J. Hölzl, Sensitec GmbH, Wetzlar.

A.Krasyuk, F. Wegelin, S.A. Nepijko, C.M. Schneider, H.J. Elmers, G. Schönhense. *Stroboscopic XMCD-PEEM imaging of nonlinear spin-wave excitation.* - BESSY Annual Report 2004, edited by K.Godehusen. Berlin: published by Berliner Elektronenspeicherring-Gesellschaft für Synchrotronstrahlung m.b.H. (BESSY), 416-418.

J. Maul, P. Bernhard, T. Berg, F. Wegelin, U.Ott, Ch. Sudek, H. Spiecker, G. Schönhense. *XANES-PEEM chemical imaging of sub-micron cosmic grains.* BESSY Annual Report 2005, edited by K.Godehusen. Berlin: published by Berliner Elektronenspeicherring-Gesellschaft für Synchrotronstrahlung m.b.H. (BESSY), 514-516.

H.J. Elmers, A. Krasyuk, F. Wegelin, S.A. Nepijko, G. Schönhense, M. Bolte, C.M. Schneider. *How to magnetize a soft-magnetic particle with microwaves.* - BESSY Highlights 2005, edited by H.Henneken, M.Sauerborn. Berlin: published by Berliner Elektronenspeicherring-Gesellschaft für Synchrotronstrahlung m.b.H. (BESSY), 12-13.

Others

F. Wegelin, Ch. Ziethen, G. Schönhense, M. Neuhäuser, R. Ohr, H. Hilgers. *Defektanalyse von a-C- und CNx-Schichten mittels Röntgen-Photoemissions-Elektronenmikroskopie (X-PEEM). Characterization of stoichiometric defects in diamond, a-C and CNx thin films with Soft X-ray photoelectron microscopy (X-PEEM).-* Vakuum in Forschung und Praxis Nr. **5** (2001) 287-292.

F. Wegelin. *Life on Mars. Sensoren helfen bei der Marserkundung.-* MessTec & Automation 05/2007, S. 42-43, GIT VERLAG GmbH & Co. KG, Darmstadt

7.2. Abbreviations

AES	-	Auger Electron Spectroscopy
AF	-	Anti-ferromagnet
AFCM	-	Antiferromagnetic Coupling Maximum
BESSY	-	Berliner Elektronen-Speicherring Gesellschaft für Synchrotronstrahlung
BLS	-	Brillouin Light Scattering
CPW	-	Coplanar Wave Guide
ESRF	-	European Synchrotron Radiation Facility, Grenoble
FIB	-	Focused Ion Beam
FM	-	FerroMagnet
FL	-	Free Layer
FWHM	-	Full Width at Half Maximum
GMR	-	Giant MagnetoResistive Effect
HDD	-	Hard Disc Drive
IBDS	-	Ion Beam Deposition System
LEED	-	Low Energy Electron Diffraction
MFM	-	Magnetic Force Microscopy
ML	-	MonoLayer
MOKE	-	Magneto-Optical Kerr Effect
MRAM	-	Magnetic Random Access Memory
MS	-	MacroSpin
NM	-	Non-magnetic Layer
NMP	-	N-Methyl-2-pyrrolidinone
OOMMF	-	Object Oriented Micromagnetic Frame
PEEM	-	Photo Emission Electron Microscopy
PVD	-	Plasma Vapour Deposition
Py	-	Permalloy
RF	-	Radio Frequency
SEM	-	Scanning Electron Microscopy
SEMPA	-	Scanning Electron Microscopy with Polarization Analysis
SNR	-	Signal to Noise Ratio
SP-STM	-	Spin Polarized Scanning Tunneling Microscopy
STM	-	Scanning Tunnelling Microscopy
TEM	-	Transmission Electron Microscopy
TEM Wave	-	Transversal ElectroMagnetic Wave
TMR	-	Tunnel MagnetoResistive Effect
TR-PEEM	-	Time Resolved - Photo Emission Electron Microscopy
UHV	-	Ultra High Vacuum
VSM	-	Vibrating Sample Magnetometer
XMCD	-	X-ray Magnetic Circular Dichroism
X-PEEM	-	X-ray Photo Emission Electron Microscopy

7.3. References

[1] J. Raabe, C. Quitmann, C. H. Back, F. Nolting, S. Johnson, and C. Buehler, Phys. Rev. Lett. **94**, 217204 (2005).

[2] K. Perzlmaier, M. Buess, C. H. Back, V. E. Demidov, B. Hillebrands, and S. O. Demokritov, Phys. Rev. Lett. **94**, 057202 (2005).

[3] S.-B. Choe, Y. Acremann, A. Scholl, A. Bauer, A. Doran, J. Stöhr, and H. A. Padmore, Science **304**, 420 (2004).

[4] H. Stoll et al., Appl. Phys. Lett. **84**, 3328 (2004).

[5] J. P. Park, P. Eames, D. M. Engebretson, J. Berezovsky, and P. A. Crowell, Phys. Rev. B **67**, 020403(R) (2003).

[6] B. C. Choi, M. Belov, W. K. Hiebert, G. E. Ballentine, and M. R. Freeman, Phys. Rev. Lett. **86**, 728 (2001).

[7] E. B. Myers, D. C. Ralph, J. A. Katine, R. N. Louie, and R. A. Buhrman, Science **285**, 867 (1999).

[8] W. Weber, S. Riesen, and H. C. Siegmann, Science **291**, 1015 (2001).

[9] S. O. Demokritov, B. Hillebrands, and A. N. Slavin, Phys. Rep. **348**, 441 (2001).

[10] B. Hillebrands, K. Ounadjela (Eds.), *Spin Dynamics in Confined Magnetic Structures I II and III*, Springer, Berlin, 2002, and 2004.

[11] O.N. Martyanov, V.F. Yudanov, R.N. Lee, S.A. Nepijko, H.J. Elmers, C.M. Schneider, G. Schönhense, Appl. Phys. A **81**, 679 (2005).

[12] P. Martín Pimentel, B. Leven, H. Grimm, B. Hillebrands, J. Appl. Phys. **102**, 063913 (2007)

[13] A. Krasyuk, A. Oelsner, S.A. Nepijko, A. Kuksov, C.M. Schneider, G. Schönhense, Appl. Phys. A **76**, 863 (2003).

[14] H. Stoll, A. Puzic, B.v. Waeyenberge, P. Fischer, J. Raabe, M. Buess, T. Haug, R. Höllinger, C. Back, D. Weiss, G. Denbeaux, Appl. Phys. Lett. **84**, 3328 (2004).

[15] D. Chumakov, J. McCord, R. Schäfer, L. Schultz, H. Vinzelberg, R. Kaltofen, I. Mönch, et al., Phys. Rev. B **71**, 014410 (2005).

[16] G. Schönhense, H.J. Elmers, A. Krasyuk, F. Wegelin, S.A. Nepijko, A. Oelsner, C.M. Schneider, Nucl. Instr. Meth. B **246** (2006) 1.

[17] P. Grünberg, R. Schreiber, Y. Pang, M. B. Brodsky, and H. Sowers, Phys. Rev. Lett. **57**, 2442 (1986).

[18] M. N. Baibich, J. M. Broto, A. Fert, F. Nguyen Van Dau, F. Petroff, P. Etienne, G. Creuzet, A. Friederich, and J. Chazelas, Phys. Rev. Lett. **61**, 2472 (1988).

[19] G. Binasch, P. Grünberg, F. Saurenbach, and W. Zinn, Phys. Rev.B **39**, 4828 (1989).

[20] S. S. P. Parkin, Phys. Rev. Let. **71**, 10 (1993)

[21] R. Coehoorn, in *Handbook of Magnetic Materials*, edited by K. H. J. Buschow (Elsevier Science, Amsterdam, 2003), Vol. 15.

[22] C. M. Schneider, O. de Haas, U. Muschiol, N. Cramer, A. Ölsner, M. Klais, O. Schmidt, G.H. Fecher, W. Jark, G. Schönhense, J. Magn. Magn. Mater. **233**,14–20 (2001)

[23] J. Stöhr; J. Magn. Magn. Mater. **200**, 470-497 (1999)

[24] A. Krasyuk, F. Wegelin, S.A. Nepijko, H.J. Elmers, G. Schönhense, M. Bolte, C.M. Schneider, Phys. Rev. Lett. **95**, 207201 (2005).

[25] J. Vogel, W. Kuch, M. Bonfim, J. Camarero, Y. Pennec, F. Offi, K. Fukumoto, J. Kirschner, A. Fontaine, S. Pizzini, Appl. Phys. Lett. **82**, 2299 (2003).

[26] A. Hubert, R. Schäfer, *Magnetic Domains: The Analysis of Magnetic Microstructures*, Springer, Berlin, 1998.

[27] Y. Zheng J.-G. Zhu, J. Appl. Phys., **81**, 8 (1997)

[28] J. Torres-Heredia, F. López-Urías, E. Muñoz-Sandoval, Journal of Magnetism and Magnetic Materials **305**, 133–140 (2006).

[29] M. Kläui, C. A. F. Vaz, J. A. C. Bland, T. L. Monchesky, J. Unguris, T. L. Monchesky, J. Unguris, S. Cherifi, S. Heun, and A. Locatelli, L. J. Heyderman, Z. Cui, Phys. Rev. B **68**, 134426 (2003)

[30] L. Landau and E. Lifshitz, Phys. Z. Sowjetunion **8**, 153 (1935); T. L. Gilbert, Phys. Rev. **100**, 1243 (1955)

[31] B. Hillebrands: *Progress in multipass tandem Fabry-Pérot interferometry*, Rev. Sci. Instrum. **70**, 1589 (1999)

[32] Z. Celinski et.al.: *Using ferromagnetic resonance to measure the magnetic moments of ultrathin films*, J. Magn. Magn. Mater. **166**, 6 (1997)

[33] S. E. Russek, S. Kaka, IEEE Transaction on Magnetics **36**, 5 (2000)

[34] C. Kittel, Einführung in die Festkörperphysik, 14. überarbeitete u. erweiterte Auflage, Oldenbourg Wissenschaftsverlag, München 2006.

[35] B. Dieny, V.S. Speriosu, B.A. Gurney, S.S.P. Parkin, D.R. Wilhoit, K.P. Roche, S. Metin, D.T. Peterson and S. Nadimi, J. Magn. Magn. Mater. **93**, 101-104 (1991).

[36] H. W. Schumacher, C. Chappert, P. Crozat, R. C. Sousa, P. P. Freitas, J. Miltat, J. Fassbender, and B. Hillebrands, Phys. Rev. Lett. **90**, 017201 (2003).

[37] Y. Pennec, J. Camarero, J. C. Toussaint, S. Pizzini, M. Bonfim, F.Petroff, W. Kuch, F. Offi, K. Fukumoto, F. Nguyen Van Dau, and J. Vogel, Phys. Rev. B **69**, 180402(R) (2004).

[38] R. Schäfer, R. Urban, D. Ullmann, H. L. Meyerheim, B. Heinrich, L. Schultz, and J. Kirschner, Phys. Rev. B **65**, 144405 (2002).

[39] H. W. Schumacher, C. Chappert, P. Crozat, R. C. Sousa, P. P. Freitas, and M. Bauer, Appl. Phys. Lett. **80**, 3781 (2002).

[40] M. C. Weber, H. Nembach, B. Hillebrands, and J. Fassbender, J. Appl. Phys. **97**, 10A701 (2005).

[41] L. Thomas, M. G. Samant, and S. S. P. Parkin, Phys. Rev. Lett. **84**, 1816 (2000).

[42] S. E. Russek, R. D. McMichael, M. J. Donahue, and S. Kaka, *Spin Dynamics in Confined Magnetic Structures*, Topics in Applied Physics Vol. **87** (Springer, Berlin, 2003), p. 93.

[43] L. Lagae, R. Wirix-Speetjens, W. Eyckmans, S. Borghs, and J. De Boeck, J. Magn. Magn. Mater. **286**, 291 (2005).

[44] Y. Tserkovnyak, A. Brataas, G. E. W. Bauer, and B. I. Halperin, Rev. Mod. Phys. **77**, 1375 (2005).

[45] G. Eilers, M. Lüttic, and M. Münzenberg, Phys. Rev. B **74**, 054411 (2006).

[46] W. K. Hiebert, G. E. Ballentine, L. Lagae, R. W. Hunt, and M. R. Freeman, J. Appl. Phys. **92**, 392 (2002).

[47] B. Hillebrands and J. Fassbender, Nature (London) **418**, 493 (2002).

[48] G. Schönhense, H. J. Elmers, S. A. Nepijko, C. M. Schneider, *Time-Resolved Photoemission Electron Microscopy*, in: *Advances in Imaging and Electron Physics*, (ed.) P. W. Hawkes, Vol. **142**, (Academic Press, Amsterdam-Boston-Heidelberg-London-New York-Oxford-Paris-San Diego-San Francisco-Singapore-Sydney-Tokio, 2004), pp. 159-326.

[49] Feikes, K. Holldack, P. Kuske, G. Wustefeld, Proceedings of EPAC 2004, Lucern Switzerland, p. 1954 and <www.BESSY.de>.

[50] F. Wegelin, A. Krasyuk, D. A. Valdaitsev, S. A. Nepijko, H. J. Elmers, G. Schönhense, C. M. Schneider, Phys. Status Solidi B, 1-7 (2009)

[51] A. Krasyuk, A. Oelsner, S. A. Nepijko, N. N. Sedov, A. Kuksov, C. M. Schneider, and G. Schönhense, Appl. Phys. A **79**, 1925 (2004).

[52] D. Neeb, A. Krasyuk, A. Oelsner, S. A. Nepijko, H. J. Elmers, A. Kuksov, C. M. Schneider, G. Schönhense, J. Phys.: Condens. Matter **17**, S1381 (2005).

[53] D.A. Valdaitsev, A. Kukunin, J. Prokop, H.J. Elmers, G. Schönhense, Appl. Phys. A **80**, 731-734 (2005).

[54] http://www.ansoft.com

[55] B. C. Dodrill, *Magnetic Media. Measurements with a VSM*, Lake Shore Cryotronics, www.lakeshore.com

[56] F. Wegelin, D. Valdaitsev, A. Krasyuk, S. A. Nepijko, G. Schönhense, H. J. Elmers, I. Krug, C. M. Schneider, Phys.Rev.B, **76**, 13 (2007).

[57] J. Malzbender, M. Przybylski, J. Giergiel, J. Kirschner: Surf. Sci. 414, 187 (1998)

[58] S. Murphy, D. Mac Mathuna, G. Mariotto, I.V. Shvets: Phys. Rev. B **66**, 195 417 (2002).

[59] H. Fritzsche, J. Kohlhepp, U. Gradmann, Phys. Rev. **51**, 15 933 (1995).

[60] J. Prokop, A. Kukunin, M. Pratzer, H.J. Elmers, J. Magn. Magn. Mater. 265, 60 (2003).

[61] R.M. Osgood III, B.M. Clemens, R.L. White, Phys. Rev. **55**, 8990 (1997).

[62] R.M. Osgood III, S.D. Bader, B.M. Clemens, R.L. White, H. Matsuyama, J. Magn. Magn. Mater. **182**, 297 (1998).

[63] T. Yaguchi, T. Kamino, M. Sasaki, G. Barbezat, R. Urao, Microscopy and Microanalysis **6**, 218–223 (2000).

[64] J. Fassbender, J. McCord, Journal of Magnetism and Magnetic Materials **320,** 579–596 (2008).

[65] V. Demidov, D. Kohlin, S. Demokritov, B. Hillebrands, F. Wegelin, J. Marien, Appl. Phys. Lett. **84**, 15 (2004).

[66] FEI Company, System Specification XL860, PN 24906-A, 1998.

[67] FEI Company, Hardware and Theory, PN 25418-A, 1998

[68] H.R. Kaufmann, J.M.E. Harper, J.J. Coumo, J. Vac. Sci. Technol. **21**, 764 (1982).

[69] A. Krasyuk, *Entwicklung der zeitaufgelösten Photoemissions-Elektronenmikroskopie für die Untersuchung der Magnetisierungsdynamik von mikrostrukturierten magnetischen Schichten*. Dissertation, Fachbereich für Physik, Johannes Gutenberg-Universität, Mainz, 2005.

[70] F. Wegelin, A. Krasyuk, H. J. Elmers, S. A. Nepijko, C. M. Schneider, G. Schönhense, Surf. Sci., **601**, 20 (2007).

[71] Parameters used for the simulations were $A = 1.3 \times 10^{-11}$ J/m, $M_s = 10.4 \times 10^5$ kA/m, $\alpha = 0.01$ and $H_y = 0.6$ mT. The particle size and $H_x(t)$ are similar to those in the experiment: the cell size is 20 nm (http://math.nist.gov/oommf/).

[72] H.W. Schumacher, C. Chappert, P. Crozat, R. C. Sousa, P. P. Freitas, J. Miltat, J. Ferré, IEEE Transactions on Magnetics **38**, 5 (2002).

[73] http://math.nist.gov/oommf/

[74] D. Atkinson, D.A. Allwood, G. Xiong, M.D. Cooke, C.C. Faulkner, R.P. Cowburn, Nat. Mater **2**, 85 (2003).

[75] K. Fukumoto, W. Kuch, J. Vogel, F. Romanens, S. Pizzini, J. Camarero, M. Bonfim, J. Kirschner, Phys. Rev. Lett. **96**, 097204 (2006).

[76] R. Varga, A. Zhukov, J.M. Blanco, M. Ipatov, V. Zhukova, J. Gonzalez, P. Vojtaník, Phys. Rev. B **74**, 212405 (2006).

[77] M. Bailleul, D. Olligs, C. Fermon, Appl. Phys. Lett. **83,** 972 (2003).

[78] R. Hertel, W. Wulfhekel, J. Kirschner, Phys. Rev. Lett. **93**, 257202 (2004)

I want morebooks!

Buy your books fast and straightforward online - at one of the world's fastest growing online book stores! Environmentally sound due to Print-on-Demand technologies.

Buy your books online at
www.get-morebooks.com

Kaufen Sie Ihre Bücher schnell und unkompliziert online – auf einer der am schnellsten wachsenden Buchhandelsplattformen weltweit!
Dank Print-On-Demand umwelt- und ressourcenschonend produziert.

Bücher schneller online kaufen
www.morebooks.de

OmniScriptum Marketing DEU GmbH
Heinrich-Böcking-Str. 6-8
D - 66121 Saarbrücken
Telefax: +49 681 93 81 567-9

info@omniscriptum.com
www.omniscriptum.com

Printed by Books on Demand GmbH, Norderstedt / Germany